The Luminiferous Aether
primary substance of the universe

Michael Heffron

ISBN: 9798869807984

The author thanks the combined Department of Mathematics and Computer Science at Hobart and William Smith Colleges for their Mandelbrot Viewer (https://math.hws.edu/eck/js/mandelbrot/MB-info.html). Appendix B lists the viewer settings used to develop the cover image for this book.

The author thanks https://editor.codecogs.com/ for their exceptional LaTeX to SVG conversion tool, used to produce all but the simplest equations in this book.

The author thanks Creative Commons (creativecommons.org) for the use of their public domain photographs, as cited.

The author also thanks all of the brave souls who agreed to review this book prior to publication.

DEDICATION

To all of my readers—I hope you will seek to know
the truth, and that the truth will set you free!

Table of Contents

PREFACE

Just over a century ago, many prominent scientists believed that luminiferous aether must convey light waves much like air conveys sound waves.[1] Those scientists were right! The next few chapters of this book explain how to examine our reality in ways that reveal the most important characteristics of the aether, and how those characteristics produce mass, energy, and surface tension. Subsequent chapters then explain how the surface tension of black holes produces gravity, electricity, magnetism, and light.

Readers won't need to understand math to understand the topics discussed in this book; however, this book does include math for those who prefer to verify the calculations. For those readers who hate math, this book puts boxes around all mathematical explanations, with their backgrounds shaded light gray to indicate material that is safe to skip by those who are not interested in the math.

> Those who love math should read all of the paragraphs, including those surrounded by boxes with light gray backgrounds (like this one). Readers are free to use (and will likely gain valuable insight from) calculus, discrete math, vector math, and much more; however, fairly simple algebra is all that is necessary to understand how the luminiferous aether produces the universe as we know it.

There are sentences within this book that may mention that something is this "times that" or "per that" (the later meaning "divided by that"). Whenever such statements become so similar to math as to be irksome, the author urges readers who hate math to ignore those terms and just interpret them as "this and that somehow make something."

---

1   Luminiferous means "light-bearing."

Billions and billions

In episode 7 (*The Backbone of Night*) of the 1980 television mini-series *Cosmos*, American astronomer Carl Sagan famously said "There are in fact one hundred billion other galaxies, each of which contains something like one hundred billion stars."

One hundred billion is a one followed by eleven zeros. Spreadsheet notation expresses that as 1e11, where the e11 means to move the decimal place (assumed when absent) eleven digits to the right, so that 1e11 represents the number 100,000,000,000.

Looking inward produces similar unimaginable numbers. There are about 1e22 atoms in two drops of water. That is about the number of stars in the universe.[2]

Spreadsheet notation is equally good for expressing very small numbers, such as one millionth (1e-6) of a second, where e-6 means to move the decimal place six digits to the left. Thus, one millionth of a second is 0.000001 second.

It can be difficult to discuss such large or small numbers, so it is common to attach prefixes to numbers, such as giga (G, meaning billion) or micro (u or μ, meaning millionth). One millionth of a second is one microsecond (1 μs, often expressed as 1 μS), and a galaxy consists of about 100 gigastars (100 G*).

Metric system

Use of the metric system helps engineers and scientists to avoid potentially serious conversion errors.[3] This book uses metric units for that very reason. To further reduce conversion errors, the Frink programming language includes units of measure in its calculations and is able to perform many conversions automatically. Appendix A includes Frink software for the most important equations found in this book to assist anyone who wishes to verify results, convert to different units, or perform further analysis.

For those readers who are unfamiliar with the metric system, the following are some useful guidelines: Seconds are the normal units of time to which you are accustomed. Kilograms are about 2.2 pounds (35.3 ounces). Meters are just over a yard (3.3 feet, 39.37 inches). Kilometers (1000 meters) are about 0.62 mile… and there is also your introduction to why conversion errors occur.

---

2 1e11 galaxies times 1e11 stars per galaxy is 1e22 stars.
3 Système International (SI) d'Unités (French for International System of Units) is a modern extension of the metric system.

Spelling, ambiguity, and morphing words

Throughout the decades, certain words have developed alternative spellings. An example is the author's use of the "archaic" spelling of the word "aether" to avoid any possibility of confusion with modern chemical ether.

This book will demonstrate that things we call "constant" are actually variable but stable. Consequently, always interpret the word "constant" (whether in quotation marks or not) to mean extremely stable or appearing not to vary. Before the end of this book, readers should realize that nothing in our reality is truly constant–not even the "speed" of light. Because English is an inherently ambiguous language, this book supplements its carefully chosen words with many analogies that attempt to clarify important points.

Language often morphs as humanity advances. The author will clarify significant morphing words as they occur in the text.

Physics

Although it may seem somewhat onerous to newcomers, physics is merely the study of physical attributes (such as color, mass, size, temperature, velocity, and much more). The "laws" of physics describe how things behave, from the astronomical down to the subatomic level.

Minutiae can easily overwhelm a reader's ability to comprehend the major concepts presented in this book. To emphasize each topic at hand, the author often very deliberately avoids minutiae. Although accurate, some descriptions are deliberately imprecise to focus readers on fundamental concepts. Due to the author's intentional imprecision and avoidance of minutiae, some readers may wish to use greater precision and/or independently research the many interesting nuances of real world situations.

At times it has been very challenging to be imprecisely accurate, so as to convey certain concepts without becoming mired in minutiae that detracts from the topic. There are many wonderful textbooks and other resources to help interested readers delve into the minutiae. For some topics, independent research can be so beneficial that the author will explicitly recommend it for those readers who may be interested.

Atoms & electrons

This book expresses many topics from the viewpoint that atoms exist–not to endorse that viewpoint, but because that view often most accurately and simply explains existing data and observations. Does that mean the view is correct? Perhaps not! That is why the examination of what is not quite right can often lead to tremendous insight. It is important to always consider other viewpoints, form conclusions, and interpret topics in whatever way seems most logical and comprehensible.

There seems to be a widespread misconception that Nikola Tesla rejected the theory of electrons. As evidenced by many of his scientific articles, Tesla was actually the first scientist to demonstrate the existence of the electron.[4] That is important to note because the role of electrons is critically important to many of the topics discussed in this book.

History

Humanity owes a great debt of gratitude to Nikola Tesla, who was the inventor of the brushless alternating current induction motor and many other devices used by nearly every modern convenience. This book mentions many relevant historical events to emphasize how recently much of our modern technology emerged.

To an elementary school student, high school may seem a drudgerous eternity away–and a century ago may seem like ancient history; however, with advancing years comes the realization that events of a century or two ago are not so distant as they once seemed.

The author grew up in a farmhouse that was built in 1913, without electricity or plumbing (as was very common during that era). Outdoors there was an outhouse and a hand operated water pump. Before he entered kindergarten, his parents and older brother built an addition onto the house that provided plumbing, bathrooms, and electricity. Next came the space race, personal computers, "smart" phones... and now the world has become a completely different place–with current generations largely unaware of that extremely rapid transition from primitive to modern.

The rapid progress of technology during the past century unfortunately produced an unrealistic awe of science. Theoretical physics is an extremely important academic discipline; however, science went too far astray and began mistaking theory for reality. Television "programming" aggravated that mistake by popularizing and propagandizing erroneous theories. Consequently, this book often reverts back to science from past centuries–when science was less wrong than it is today (because it was based upon observations rather than theories or models).

Albert Einstein said "If we knew what it was we were doing, it would not be called research, would it?" **Science has always been wrong. Science is wrong now. Science will always be wrong.** That is the nature of science–it is a study, a quest for knowledge, an attempt to discover the truth about how things work. Each new discovery brings humanity a little closer to the truth; however, despite what this book reveals, we are still very far from understanding how bubbles and ripples in the aether became the universe as we know it.

---

4 Interested readers should refer to Tesla's actual views, such as those found at
   drnikolatesla.tumblr.com/post/190987442413/nikola-teslas-views-on-the-electron.

To avoid delving too far into potentially incorrect science, the author avoids taking readers beyond the boundaries of how the aether produces energy, mass, gravity, electricity, magnetism, and light. Regardless of how right or wrong it may be, interested readers can apply contemporary related science to understand how the aether impacts everything beyond those boundaries (and thereby possibly discover errors in that related science).

Viewpoints

In a prior book, the author described gravity, mass, and electromagnetism in terms of Kepler's Third Law and the characteristics of black holes.[5] This book extends that analysis to explain how all of that and more results directly from the aether. After a brief explanation of fundamental concepts and an extensive discussion of density and pressure, the remaining chapters describe how the aether produces the essence of our reality.

Essential to the progression of science is the ability and willingness to look at things from different viewpoints. It is then important to examine whether those viewpoints help to understand the observed behavior. **Everything we think we know is not quite right–but, if we keep seeking the truth, we will get closer to the truth with each new discovery.**

Consequently, this book advocates viewing things from the bottom up, top down, inside out, outside in, and at times leaping entirely out of the box to view things from a totally different perspective! Readers may occasionally experience the odd sensation of déjà vu as this book examines certain topics from different perspectives.

Questions

Thought-provoking questions are sprinkled throughout this book to encourage readers to think about certain topics. When too complex for mere thought, those questions appear as a "topics for further research" section at the end of many chapters.

---

5   Heffron, Michael. *Ramblings of an Old Man: about the scientific method* (self-published via Kindle Direct Publishing, 2020).

1 ESSENTIAL CONCEPTS

This chapter explains some basic concepts that are essential for fully understanding the remainder of this book. These concepts are often from historical observations, with suggestions for viewing some things differently to gain new insights. A different viewpoint often facilitates a deeper comprehension of why things behave as they do.

In keeping with the goal of science to seek and discover the truth, readers should always view things from whatever perspective provides the broadest understanding of how things interact with each other.

Velocity

Ancient humans quickly learned that hares are faster than tortoises, and panthers are faster still. During the time it took the sun to cross the sky, they learned they could walk to a distant landmark and back. Those are all examples of velocity, which is distance per time. A more modern example of velocity is meters per second.[6]

Acceleration

Acceleration is velocity that changes over time. A car departing from a stop sign is a good example of acceleration–as its velocity increases every second (meters per second per second, which is meters per second squared) until it reaches the posted speed limit. Deceleration is simply negative acceleration.

6　This book purposefully omits the distinction that a mathematical **velocity** vector is distance per time in a specific direction. Although no topic within this book requires knowledge of vector math, it can be an interesting topic for readers to pursue independently.

Acceleration describes the rate of change without regard to the initial velocity. For example, a car that is already traveling 100 kilometers per hour (kph) may accelerate 5 kph per second for each of four seconds to pass a truck at 120 kph. It may then decelerate 2 kph (accelerate −2 kph) per second for each of ten seconds to slow back down to 100 kph after passing.

Gravity

We are all acquainted with the acceleration of gravity toward the earth. Toddlers quickly learn that gravity makes them fall. Gravity also quickly accelerates any food they push over the edge of their highchair. That food falls toward the floor at a rate of about 10 meters per second squared (m/s^2).[7]

Objects thrown into the air quickly stop, reverse direction, and fall back toward earth. For example, a tennis ball thrown into the air at a velocity of 25 meters per second (m/s) decelerates to a velocity of 15 m/s after one second. By the end of another second, the ball slows to a velocity of 5 m/s. In about another half second, the ball stops (zero velocity) at its maximum height in the air.

Gravity then pulls the ball back to earth. After falling back for one second, the ball's velocity will be 10 m/s. At the end of another second, the ball will have a velocity of 20 m/s. About half a second later, the ball will return to earth with a velocity of 25 m/s. During the first 2.5 seconds the ball decelerates from 25 m/s to 0, and during its final 2.5 seconds the ball accelerates from 0 to 25 m/s.

Frame of reference (viewpoint)

The prior topic described gravity from the frame of reference of a bystander watching the events transpire on earth. From the viewpoint of an ant clinging onto the tennis ball, everything would appear very different.

To the ant on the ball being thrown, it would seem like the earth suddenly darted off. Then it spun around, slowed down, stopped, started drifting back, spun around some more, and suddenly slammed back into the ball with a tremendous jolt. The earth might have also bounced off the ball a few times thereafter. It can be enlightening to visualize (or view) many phenomena from such alternative frames of reference.

Weight, mass, and force

Our prehistoric ancestors discovered that big rocks are heavier than small stones. They learned how to weigh things on primitive scales. Ancient shipbuilders learned that enormously heavy boats can float on the water as if weightless.

---

7   As warned about in the preface, this is a case of intentional slight imprecision (rounding the 9.8 m/s^2 acceleration of gravity to "about" 10 m/s^2) to avoid having minutiae needlessly detract from the discussion. To focus readers on the concept of gravity, this topic also ignores the negligible air resistance for this example.

Mass (kilograms, kg) is the amount of matter in a body. Scientists ultimately realized that force (newtons, N) is mass times acceleration, and weight is the force that results from gravity accelerating mass toward the earth.

To reduce possible confusion about mass versus weight, the metric system meticulously distinguishes between weight and mass. That is important because 100 kg of mass weighs 100 N under the influence of normal gravity, but that same 100 kg of mass weighs 0 N (is "weightless") when orbiting the earth.

We are so accustomed to measuring things relative to our surroundings that we often don't even pause to realize that all measurements are relative to some frame of reference, just as weight is relative to gravity. More than many other topics, mass versus weight demonstrates the extreme importance of understanding the frame of reference. Consequently, the author may emphasize important but non-obvious frames of reference.

Momentum and conservation thereof

Mass times velocity is momentum (kilogram·meters per second, kg·m/s). Even a stationary object has momentum, albeit that momentum is zero from whatever frame of reference considers it to be stationary.

When objects collide, they redistribute their momenta among the objects (or pieces/clumps thereof). Collisions conserve momentum, which means the sum of all momenta involved in the collision remains constant.

The equation

Eq 1-1 $$m_1 u_1 + \cdots + m_n u_n = m_1 v_1 + \cdots + m_n v_n$$

expresses the conservation of momentum, where $m_\#$ represents the mass of object #, $u_\#$ represents the pre-collision velocity of object #, $v_\#$ represents the post-collision velocity of object #, n represents the number of objects involved in the collision, and ... denotes unlisted terms for objects 2 through n-1.

The popular "Newton's cradle" toy uses swinging spheres to demonstrate the principle of *conservation of momentum*.[8] Upon release of the first sphere, gravity accelerates it into the second sphere–thereby stopping the first sphere as its momentum transfers to the second sphere. With minimal movement, each adjacent sphere transfers its momentum to the next, until the last sphere swings up into the air. Just like the tennis ball that was thrown into the air, gravity decelerates the last sphere to zero velocity, then restarts the process by accelerating that sphere back toward the idle spheres. The end spheres keep

---

8   Readers who are not familiar with the "Newton's cradle" toy should take the time to look it up on Wikipedia (which has an excellent animated GIF), or YouTube (which has many excellent videos).

repeating that cycle, interacting with gravity to transfer momentum back and forth, until friction ultimately slows them to a standstill.[9] **Momentum and conservation thereof are among the most important concepts in all of physics!**

Inertia

Due to momentum, an object in motion tends to remain in motion. Similarly, an object at rest tends to remain at rest. In addition to simple (straight line) momentum, rotating or swinging objects can experience angular momentum. Newton's first law of motion (the Law of Inertia) states that every object tends to remain at constant velocity (at rest or in uniform motion), unless an external force intervenes. For example, the momentum of an object doesn't change until it interacts with another object in a way that alters their momenta.

Inertia keeps a released bowling ball rolling down the alley until it reaches the pins. Inertia also keeps the middle spheres of a Newton's cradle near zero velocity, while the acceleration of gravity overcomes and then restores the inertia of the end spheres as they swing back and forth.

Temperature

Our prehistoric ancestors understood that fire is extremely hot, summer days can be uncomfortably hot, and winter days are cold. Modern humans understand that molecules of a substance collide with each other ("vibrate") more as temperature rises. Those vibrations tend to push nearby molecules farther apart, thereby causing the substance to expand.

A typical mercury thermometer uses expansion to measure temperature—usually via a bulb filled with mercury connected to a thin cylinder. Even small expansions of the mercury easily propel it upward past the temperature markings beside the thin cylinder (which often is also a magnifying lens).

As temperature drops, molecules of a substance vibrate less. Those reduced vibrations allow nearby molecules to move closer, thereby causing the substance to contract.

The increased momentum needed to push molecules farther apart requires increased velocity. Otherwise, the molecules would keep the same spacing. To push them apart, the final velocity must be at least marginally greater than the initial velocity. That requires energy!

---

9   Air resistance also contributes minimally to stopping the oscillations. It is ironic that the study of minutiae such as friction and air resistance can be some of the most interesting aspects of physics as well as the greatest hindrance to seeing the big picture!

Energy

It is common to define energy as mass times velocity squared ($kg \cdot m^2/s^2$), and to carefully distinguish between potential energy that is available for use and kinetic energy that is "consumed" to do work (such as pushing molecules farther apart).[10] Relative to some frame of reference, the concept of energy is most useful when discussing consumption, production, or storage.

According to the First Law of Thermodynamics, **energy cannot be created nor destroyed, it can only be converted from one form to another**.[11] Although not always a practical viewpoint, it can be preferable to define energy as initial momentum times final momentum divided by the mass. That viewpoint can sometimes assist with better accounting for how *conservation of momentum* produces *conservation of energy*. It also reveals why kinetic energy is only half the magnitude of potential energy.

In whatever frame of reference views energy as consumed, there are many alternative viewpoints where that energy was produced or stored. Referring back to the gravity example, it looked like the earth did all of the work from the perspective of the ant clinging to the tennis ball; however, the person who threw the ball may have thought they did all of the work. If gravity were sapient, it might argue it did all of the work.[12] Each would be correct from their own perspective. Due to *conservation of momentum*, energy cannot be created nor destroyed–it can only be converted from one form to another, relative to a particular frame of reference.

Both momentum and energy are *always* relative, because motion is always relative! It's important to be aware of the relative and complex nature of energy, but failure to master that complexity shouldn't prevent comprehension of the topics within this book. The author encourages readers who want to know more about energy to independently research that topic.

Mass flow and surface tension

Turning on a faucet produces a flow of some volume of liquid per second, which is also a mass flow (kilograms per second, kg/s). Surface tension is mass flow *per second* (kilograms per second *per second*, kg/s^2).

Surface tension causes liquid surfaces to shrink into the minimum possible surface area. In other words, it causes the mass at the surface of the liquid to flow

---

10 Although beyond the scope of this book, note that conservation of momentum causes kinetic energy to be only half the magnitude of potential energy.

11 Interested readers should independently research thermodynamics for a comprehensive explanation of what such conversions entail.

12 People often confuse the terms sapient and sentient. Sapient beings exhibit intelligence, whereas sentient beings experience emotions and sensations. Many living creatures (such as humans and animals) are both sapient and sentient.

toward the interior of the liquid. With great care, it is often possible to fill containers slightly beyond their maximum capacity, at which point surface tension causes the liquid to bulge slightly without spilling over the edges of the container. That may be easiest to demonstrate by using an eye dropper (or a small straw) to slowly add water to the surface of a penny.

By limiting how much mass can flow toward the interior, surface tension causes droplets of water to bead up on water repellent surfaces, and enables many insects to walk on water. In addition, surface tension causes capillary action and various other interesting phenomena.

 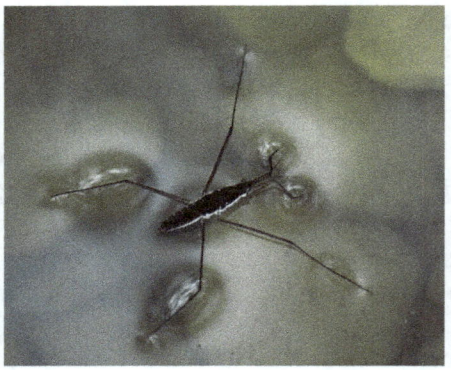

(a) water beading on a leaf[13]   (b) insect walking on water[14]

Figure 1-1. Common examples of surface tension.

In stages, future chapters reveal that **the luminiferous aether causes surface tension within black holes that produces gravity, electricity, magnetism, and light**. Interested readers will find a wealth of information if they choose to independently research the topic of surface tension; however, always keep in mind that science is not quite correct.

---

13 Photo by Michael Apel, CC BY-SA 3.0
   <http://creativecommons.org/licenses/by-sa/3.0/>, via Wikimedia Commons
14 By PD - Wikipedia english, CC BY-SA 3.0,
   https://commons.wikimedia.org/w/index.php?curid=1095618

2 DENSITY, PRESSURE, AND THE "SPEED" OF SOUND

This chapter encourages readers to gain new insights by learning to view everything differently. Even those readers who already understand density and pressure can benefit enormously from this different viewpoint.

It is common to view density as mass per volume, and pressure as force per area.[15] This chapter emphasizes the view of pressure times volume as energy.[16] Future chapters emphasize the exceptionally important view of pressure times distance as surface tension (which is also equivalent to energy per area).

It is insufficient to view pressure only as something that inflates a balloon, or density as the thing that makes a lighter-than-air balloon float or a heavier-than-air balloon sink. Pressure and density both occupy the same volume within the balloon, and are just two of many different coexisting views (frames of reference) for the same group of molecules.

The role of momentum

Relative to the frame of reference of the inflated balloon, each individual gas molecule has some mass and velocity. Thus, each molecule also has a momentum (which is equal to its mass times its velocity), and an energy (which is equal to its momentum times its velocity, or its mass times its velocity squared).

Combinations of molecules within the balloon may be part of a one-dimensional strand, a two-dimensional area, or a three-dimensional volume. As shown by Table 2–1, **all of those views of the gas molecules within the balloon simultaneously coexist and are related to each other by velocity and/or distance/dimension.**

---

15 For example, many people state tire pressure as "pounds" (of force) per square inch.
16 Even on a small scale, it becomes obvious that pressure times volume is energy. For example, when excessive inflation explodes a balloon.

Table 2-1. Important simultaneous ways to view fluid molecules.

	× velocity →		
× dimension →	mass (kg)	momentum (kg·m/s)	energy (kg·m²/s²)
	linear density (kg/m)	mass flow (kg/s)	force (kg·m/s²)
	area density (kg/m²)	viscosity (kg/m·s)	surface tension (kg/s²)
	density (kg/m³)	mass flux (kg/m²s)	pressure (kg/m·s²)

Each view (frame of reference) in Table 2–1 is the view beneath times distance (or one dimension) and/or the view above per distance (or one dimension). For example, energy (top right) is force times distance (one dimension), or surface tension times area (two dimensions), or pressure times volume (three dimensions). Similarly, pressure (bottom right) is surface tension per distance, or force per area, or energy per volume.

Each column to the left is the view per velocity and each column to the right is the view times velocity. For example, density (bottom left) is mass flux (bottom center) per velocity, and pressure (bottom right) is mass flux times velocity (or density times velocity squared).[17]

Each table entry describes a different view of the gas molecules within the balloon. Dimension and/or velocity determine(s) how each view relates to every other view. It is extremely important to understand that all of those views are valid simultaneously within the inflated balloon, which is itself an encompassing view (the volume that contains the molecules of every view).[18]

All of those views can easily be confusing! It is common to believe the views are different things because different instruments are necessary to measure mass,

---

17 Various configurations of column(s) and row(s) define many other such views. For example, acceleration is force divided by mass. It should also be obvious why mass times velocity squared is energy, or pressure per density is velocity squared (as is energy per mass).

18 To avoid becoming bogged down in minutiae, the text glosses over the fact that the cells of Table 2–1 may refer to individual molecules, or the average for a group. A one dimensional strand could be a sliver of a slice, or a path, or a stream—and an area could be a slice of a volume, or a shape therein, or its surface.

energy, force, viscosity, density, pressure, … with each instrument specifically designed to measure the fluid from an appropriate frame of reference; however, different measurement instruments notwithstanding, all of those views simultaneously coexist.

A simple example of that is a pail of water. Its molecules simultaneously have mass, volume, temperature, density, pressure, viscosity, and many other valid frames of reference (each of which requires specialized measurement instruments).

Temperature is proportional to kinetic energy

As momentum increases, temperature also increases. Mercury expands in the bulb of a thermometer as the increased momentum of its molecules provides the kinetic energy (momentum times the velocity of collisions) that pushes them farther apart. It is important to remain focused on the fact that temperature, momentum, and kinetic energy are merely different ways to view (or measure) the same molecules.

Collisions that increase velocity also increase temperature, momentum, energy, and pressure–but decrease density. Conversely, collisions that decrease velocity also decrease temperature, momentum, energy, and pressure–but increase density. That is, those collisions we view as conservation of momentum are simultaneously legitimate to view as mass, energy, mass flow, force, viscosity, surface tension, density, mass flux, pressure, and much more. It doesn't really matter whether you call those attributes, characteristics, frames of reference, measurements, or views–the significant point is that they are all different perspectives of or within the same group of molecules.

Buoyancy

Buoyancy is the tendency of a body to float when submerged in a fluid. Ancient Greek inventor and mathematician Archimedes discovered that a ship floats when the water it displaces weighs as much as the ship.

Although buoyancy is most often attributed to density, many readers may already realize that pressure is equal to the buoyant force (which keeps a ship from sinking) per submerged area. Therefore, **pressure must equally well determine buoyancy** (because density and pressure are merely different ways to view the same molecules within the volume of their shared frame of reference).[19]

The submerged area of a hull is a good example of partially-immersed buoyancy. Hot air balloons and the floating bulbs of the "Galileo thermometer" are two inventions from the 1700s that provide superior examples of completely immersed buoyancy.

---

19 Interested readers may wish to research static and dynamic fluid mechanics.

The air that we breathe

Air molecules consist of mostly nitrogen, some oxygen, water vapor, dust, and usually only traces of anything else. Inertia causes those molecules to travel at constant velocity between random collisions.

Density, pressure, and temperature are each different views of the same shared momentum. Due to the potentially differing mass of molecules, collisions alter their velocities as necessary to conserve momentum.

Sunlight or other localized heating of air may increase the kinetic energy of air molecules and thereby increase their pressure and/or decrease their density. Similarly, shadows cast by clouds or other localized cooling of air may decrease the kinetic energy of air molecules and thereby decrease their pressure and/or increase their density.

Sound and its "speed"

Our distant ancestors noticed that thunder occurs sooner following lightning as a storm approaches.[20] From that time on, the "speed" of sound became an interesting topic for research.

Sound causes lower velocity compressions and higher velocity expansions (rarefactions) in the momenta of colliding air molecules. Those may also be viewed as deviations of density and pressure. That is, sound is an oscillation between density and pressure that consists of longitudinal waves.[21]

Note that sound waves propagate, rather than actually traveling through the air. In other words, air molecules mostly just vibrate in place, thereby giving the illusion of traveling waves as those vibrations spread their energy and thereby diminish. That is why sound gets weaker with distance.

It is the disruption caused by the sound that causes waves to ripple within the air, much like ripples of water caused by tossing a pebble into a pond. Waves produce the illusion of something traveling through the medium, but those waves are just oscillations (compressions and rarefactions) of the medium that usually redistribute energy without a net movement of particles.

Waves are real, despite their "travel" being largely an illusion. For example, the crests of ocean waves have enough energy to lift massive ships; however, careful observation reveals the ship remains approximately in place horizontally, while the waves primarily move it up and down vertically. Smaller items, such as

---

20 As a rough approximation, sound waves "travel" about a kilometer per three seconds or about a mile (1.6 kilometer) per five seconds.

21 Longitudinal waves align with the direction of apparent travel. Visualize that as if someone were playing an accordion, or expanding and compressing a spring. Even better, extend that visualization to expanding and contracting three-dimensional concentric bubbles.

beach balls and driftwood, may actually float in the troughs between waves and thereby travel with the energy redistribution of the "illusion."

Isn't it interesting that the waves can be real for small items (which fit between waves) but illusory for large items (which span several waves without being anchored to those waves)? Now, imagine a boat whose pontoons ride on two adjacent waves. Consider how the pontoons may anchor the boat within the troughs of the waves or may span across the waves, depending on the orientation of the boat. Under what circumstances will the pontoons travel with the waves versus merely rising and falling?

It may be helpful to visualize air as molecules traveling effortlessly on the path of their inertia until they collide with other molecules, which then rebound off each other and potentially redistribute their momenta. Disregarding wind and minor perturbations, the square of the *average* speed of sound is the air pressure divided by its density. That velocity squared results from the mobilizing velocity of air pressure times the inhibiting velocity of air density. If necessary, refer back to Table 2–1 to clarify why pressure divided by density is velocity squared.

The propagation speed of the energy redistribution by sound waves is

Eq 2-1
$$c = \sqrt{\frac{P}{\rho}},$$

where P is the pressure of the medium (such as air) through which the sound waves propagate, and ρ (Greek letter rho) is the density of the medium. The letter c represents speed (from the Latin word *celeritas*, meaning velocity).

Although they are merely massless oscillations, sound waves can exhibit tremendous power via their influence on the pressure and density of the air. Glass broken by exceptionally close thunder is a good natural example of the extreme energy that may reside within sound waves. Due to Boyle's Law (next topic), when lightning superheats the air it produces an intense high pressure rarefaction that shoves surrounding air molecules into a high density compression, and thereby propagates high energy thunder in all directions via collisions among the molecules.

The speed of sound is not a limit but rather it is the average propagation velocity of the oscillating waves produced by collisions between the molecules. Due to conservation of momentum, some of those molecules must travel much faster (such as lighter nitrogen) or slower (such as heavier oxygen) than the average speed of sound.

In a clear demonstration that the speed of sound is not a limit, hypersonic vehicles fly through air molecules at many times the speed of sound. On very many levels, that has important implications for the speed of light in the aether.

Boyle's Law (also known as the Boyle-Mariotte Law)

In 1662, Irish physicist Robert Boyle published his research on the relationship between pressure and volume for a confined gas.[22] In 1679, French physicist Edme Mariotte independently discovered the impact of temperature upon pressure and volume.

Boyle's Law essentially states that

Eq 2-2
$$\frac{P_1 V_1}{T_1} = \frac{P_2 V_2}{T_2},$$

where P_1 is the initial pressure V_1 is the initial volume, T_1 is the initial temperature, P_2 is the final pressure, V_2 is the final volume, and T_2 is the final temperature. Consequently, when one of those variables changes, one or both of the others must also change to conserve the constant ratio.

From Table 2-1, pressure times volume is energy, so Boyle and Mariotte essentially discovered that the ratio of potential energy to kinetic energy remains constant for a confined gas. Consider the shared momentum of those gas molecules and realize that Boyle's Law results from the conservation of momentum.

As a consequence of Boyle's Law, when the temperature of the air rises, its pressure must increase (if its volume is confined) and/or its density must decrease (if its volume can vary). That means the speed of sound increases when the temperature of the air increases, and the speed of sound decreases when the temperature of the air decreases. On many levels, that also has important implications for the speed of light relative to the temperature of the aether.

Density, pressure, temperature, and volume are all very important views in this book. Also important is that (within the confined gas) pressure times volume is potential energy and temperature represents the ratio of potential energy to kinetic energy. For the ratio to remain constant, potential energy must go up as kinetic energy goes up, and potential energy must go down as kinetic energy goes down.

Boyle's Law only applies to "ideal" gasses; however, readers who wish to explore the nuances (and minutiae) should independently research the various "real gas" laws (and similar laws for liquids and solids) to understand their differences and similarities.

---

22 Boyle, Robert. *A Defence of the Doctrine Touching the Spring and Weight of the Air* (London: Thomas Robinson, 1662).

3 THE LUMINIFEROUS AETHER

Ancient philosophers believed *quintessence* permeated all of nature and was *the substance that composed celestial bodies*. By the late 1800s, numerous scientific discoveries morphed that belief into the postulation that *luminiferous aether* must permeate all of space to provide *the medium by which light propagates*.

Just over a century ago, a majority of scientists still believed light waves must propagate via an aether that fills space, much like sound waves propagate via the air that fills our atmosphere. The master engineer behind all modern conveniences, Nikola Tesla, was one of those scientists who believed in the aether.[23]

One of the greatest mistakes anyone can make is to assume something doesn't exist because it couldn't be found. When the Michelson-Morley experiment of 1887 failed to detect "The Luminiferous Aether" of which De Volson Wood wrote in 1886, many other theories and experiments subsequently arose, and the theory of aether gradually fell out of favor; however, we should all ask the question "If it were possible, what would it look like?" much more often![24]

Density and pressure of the aether

Despite not being found, if aether actually does exist, it should have characteristics very similar to air–since the behavior of air is what prompted the conjecture that aether must exist in the first place. That means aether should be a fluid (gas and/or liquid) that consists of colliding particles whose momentum

---

23 The April 8, 1934 edition of the New York Times used the contemporary spelling of aether when they quoted Nikola Tesla as saying "Light cannot be anything else but a longitudinal disturbance in the ether, involving alternate compressions and rarefactions. In other words, light can be nothing else than a sound wave in the ether."
24 Asked by "Rudy" in the 1993 true story motion picture named for him.

produces density and pressure.[25] We know that sound waves are longitudinal and propagate at a speed determined by the pressure and density of the air, so that should also be true of light propagating through the aether.

Conveniently, c has already come to represent the speed of light. If we let P represent the pressure of the aether, and ρ represent the density of the aether– then the equation for the speed of sound (Eq 2-1) should work equally well to find the speed of light waves in the aether.[26]

Eq 3-1
$$c = \sqrt{\frac{P}{\rho}}.$$

Rearranging Eq 3–1 to solve for pressure,

Eq 3-2 $$P = \rho c^2$$

is a general equation that applies everywhere throughout the vast expanse of the aether. Thus, it is reasonable to view the pressure of the aether as whatever constitutes its density colliding with itself at the speed of light.

If aether truly composes all celestial bodies, at a minimum, aether must be at least as dense as a black hole. Suppose the density of the aether exactly equals that of a black hole (1.78449e16 kilograms per cubic meter).[27] If light waves propagate through aether like sound waves propagate through air, the pressure of the aether must then be 1.60382e33 newtons per square meter (equivalent to 1.60382e33 joules per cubic meter).[28] This and future chapters will demonstrate those are the proper values for the density and pressure of the aether (based upon currently defined "constants").

Since the molecules of air determine its properties, it is reasonable to speculate the substance of the aether similarly determines its properties. Whether actual, quantum fluctuations, virtual particles, or something else–whatever

---

25 From Table 2-1, observe that momentum per volume is mass flux, density is mass flux per velocity, and pressure is mass flux times velocity.

26 International agreement defines the speed of light as the constant 2.99792458e8 m/s *in vacuum* (which the author interprets to mean *in aether* that is devoid of obstructions). Thus, the nations decree an improbable constant ratio for the pressure of the aether per its density.

27 Heffron, Michael. *Ramblings of an Old Man: about the scientific method* (self-published via Kindle Direct Publishing, 2020). Chapter 7 (pages 49-50) advocates that precise density for black holes.

28 Using Eq 3-2, pressure of the aether must be its density times the speed of light squared. As shown by Table 2–1, newtons (force) per square meter is the same kilograms per meter second squared (pressure) as is joules (energy) per cubic meter.

constitutes the aether seems to behave like a fluid composed of minuscule neutral particles of matter.

Mass and energy due to aether

Just as for air, the mass of aether must be its density times its volume. Similarly, its pressure times its volume must be potential energy when contained/stored or kinetic energy when released/used. BlackHole.frink in Appendix A demonstrates those views (from Table 2–1), which simultaneously describe their shared momentum from different frames of reference.

Let § represent a shape factor for subatomic particles composed of the aether. It may be best to visualize those shapes as amorphous blobs; however, since surface tension often causes the particles to behave like spheres, substituting the scalar multiplier $\frac{4}{3}\pi$ (to find the volume of a sphere) for § is often a very good approximation. For example, $\S R^3 \approx \frac{4}{3}\pi R^3$, where R is the radius of the sphere. The volume of an ellipsoid $\left(\frac{4}{3}\pi ABC\right)$ is also a good approximation– where A, B, and C are the lengths of the semi-axes of the ellipsoid (spherical when $A=B=C$).

For any given particle of radius R, density (ρ) of the aether determines its mass,

Eq 3-3 $$m = \rho \S R^3.$$

For any given particle of radius R, pressure (P) of the aether determines the energy of its mass at the speed of light,

Eq 3-4 $$E = P \S R^3.$$

Some might call that "relativistic" energy (relative to the aether).

Although it is correct to say mass times the speed of light squared is energy, in terms of aether it is even more correct to say energy is the pressure of the aether times the volume of its mass. It may be helpful to view aether as numerous small frictionless spheroids obeying the laws of inertia and momentum to somehow flow and produce larger particles. While contained within their volume, those particles have a mass that is equal to their density times their volume. When that volume releases its particles, they escape with a potential energy that is equal to their pressure times their former volume.

Infinitesimal whirls of prodigious velocity

Nikola Tesla believed that matter is high velocity swirls of aether. In his own more eloquent words, "The primary substance, thrown into infinitesimal whirls of prodigious velocity, becomes gross matter; the force subsiding, the motion ceases and matter disappears, reverting to the primary substance."[29] Those "whirls" are the aether swirling around within the boundaries of a black hole, and the "prodigious velocity" is the speed of light.

Pressure times volume is energy. Density times volume is mass. Thus, multiplying each side of Eq 3-2 ($P = \rho c^2$) by a specific volume produces Einstein's much more familiar energy equation,

Eq 3-5 $$E = mc^2,$$

where E is the energy released, m is the mass of a particle, and c is the speed of light. That is the same energy as Eq 3-4, for the luminiferous radius (R) of a particle of mass m.

Rearranging Eq 3-3 to find the radius (R) from the mass (m) yields

Eq 3-6 $$R = \sqrt[3]{\frac{m}{\rho \S}},$$

where ρ is the density of the aether. The aethereal view is that R is the luminiferous radius for the sphere that contains the specified whirling mass, composed of the density of the aether.

Whatever aether particles are, it appears that the Law of Inertia causes their momenta to remain on unaltered trajectories between "elastic" collisions, and then conserve those momenta as they depart those collisions on different trajectories–as if they are part of a three dimensional cosmic "Newton's cradle" toy.

Topics for further research

Could the pressure of the aether behave like a gas, while the density of the aether behaves like a liquid? On a cosmic scale, could the pressure and temperature of the aether be a "triple point" at which the aether can simultaneously exist as a

---

29 Bearden, T. E. *Solutions to Tesla's Secrets and the Soviet Tesla Weapons with Reference Articles for Solutions to Tesla's Secrets* (Millbrae, CA: Tesla Book Company, 1981). Page 91 contains the quotation at the end of the article *Man's Greatest Achievement* by Nikola Tesla, from the July 6, 1930 edition of *New York American*.

gas, liquid, and solid—much like the triple point of hydrogen (13.84K@7.04kPa) or water (273.16K@611.657Pa)?[30]

---

30 Pascal (Pa) is the SI unit for pressure.

4 HOW AETHER PRODUCES GRAVITY

In the early 1600s, German astronomer Johannes Kepler derived laws of planetary motion–based upon precise astronomical observations of the planets made by the Danish astronomer Tycho Brahe in the 1500s. A slight rearrangement of Kepler's Third Law of planetary motion reveals that the velocity squared of a planet/satellite times the semi-major axis of its orbit is the constant gravitational parameter of the body/planet it orbits.[31]

To focus on important concepts, rather than the many distractions associated with elliptical orbits, this book will discuss all orbits as though they are perfectly circular–where, based upon Kepler's Third Law, **the gravitational parameter is the velocity squared of a satellite times the radius of its orbit.**[32]

Astronomer's typically use the Greek letter Mu (μ) to represent the "standard" gravitational parameter; however, due to problems with the astronomical definition, and to avoid confusion with the magnetic constant (μ_0) presented in the next chapter, this book substitutes the ancient Greek letter Koppa ($\varkappa$) to define the gravitational parameter based upon Kepler's Third Law. Thus,

Eq 4-1 $$\varkappa = v^2 r,$$

where v is the velocity of a satellite, and r is the radius of its orbit.

---

31 For earth, that gravitational parameter is 3.986004418e14 m³/s².
32 Heffron, Michael. *Ramblings of an Old Man: about the scientific method* (self-published via Kindle Direct Publishing, 2020). Chapter 5 (page 28) expresses the gravitational parameter as instantaneous velocity squared times instantaneous radius of orbit. As an alternative, note that it is easy to substitute the average velocity squared times the semi-major axis to account for elliptical orbits.

Black holes

British theoretical physicist Stephen Hawking and numerous other scientists theorized that black holes are bodies so massive that not even light can escape their gravitational pull (hence the name "black hole"). As should always be expected from science, that theory is not quite correct (despite Hawking's ingenious refinements).

Kepler's Third Law _must_ result from a black hole within the core of _every_ body.[33] Every object in the universe that has mass is a black hole, which must have a radius close to its core where the orbital velocity reaches the speed of light. That infinitesimal whirl of prodigious velocity encapsulates the density of the aether into mass (momentum per velocity) and pressure into potential energy (momentum times velocity). Because they have mass, even subatomic particles must be black holes. That is, every particle of our being consists of black holes!

Clearly, black holes are not always the voracious monsters we were led to believe. They appear to be bubbles in our reality where swirling encapsulates the aether, but they don't seem to aggressively gobble up everything around them as theorized. Could black holes be akin to Mandelbrot sets (the black blobs on the cover of this book) whose matter doesn't diverge to infinity?

As to be expected from science, this and future chapters will reveal that nearly everything we believed about black holes was not quite correct. Always remain aware of the fact that science is a flawed but noble quest to seek the truth!

Luminiferous radius

In 1916, German astronomer Karl Schwarzschild used Sir Isaac Newton's gravitational constant to solve Einstein's field equations, and thereby determine the radius of the event horizon (a theoretical boundary beyond which not even light can escape from the gravitational pull of any given black hole). When the velocity of an orbiting satellite reaches the speed of light, it is at the gravitational radius of the black hole it orbits. All of those radii rely on Newton's not quite correct gravitational constant.[34] Consequently, this book instead declares **the _luminiferous radius_ of a body is that distance at which an unobstructed satellite orbits the body at the speed of light due to Kepler's Third Law.**

---

33 Heffron, Michael. _Ramblings of an Old Man: about the scientific method_ (self-published via Kindle Direct Publishing, 2020). Chapter 6 explains in more detail why every object with a gravitational parameter must have a black hole at its core.

34 Heffron, Michael. _Ramblings of an Old Man: about the scientific method_ (self-published via Kindle Direct Publishing, 2020). Chapter 6 (page 43) explains that the formal definition of the _Schwarzschild radius_ isn't quite right, because it relies upon Newton's not quite correct Law of Gravitation (the error of which Chapter 5 of that book explains in great detail).

Thus, the luminiferous radius equals the gravitational parameter of the body divided by the speed of light squared.[35] That means the gravitational parameter is equal to the luminiferous radius times the speed of light squared.

The gravitational parameter ($\varkappa$) of any given black hole is

Eq 4-2 $$\varkappa = \frac{PR_L}{\rho},$$

where P is the pressure of the aether, and ρ is the density of the aether. That is equivalent to Eq 4-1 for a satellite orbiting the luminiferous radius (R_L) at the speed of light.

Surface tension

As indicated by Table 2–1, surface tension is pressure times distance. **The surface tension of a black hole is its luminiferous radius times the pressure of the aether or its gravitational parameter times the density of the aether.**

Multiplying both sides of Eq 4-2 by the density of the aether makes it clear that surface tension (γ) is

Eq 4-3 $$\gamma = PR_L = \rho\varkappa,$$

where P is the pressure of the aether, R_L is the luminiferous radius of the black hole, ρ is the density of the aether, and $\varkappa$ is the gravitational parameter of the black hole.

Having two ways of finding the surface tension of a black hole indicates that the surface tension due to pressure of the aether is in equilibrium with the surface tension due to density of the aether. **The gravitational parameter of a black hole is its surface tension per density of the aether, while its luminiferous radius is its surface tension per pressure of the aether.**

Swapping the ends of Eq 4-3 and dividing by the density of the aether (ρ) reveals that the gravitational parameter ($\varkappa$) of any given black hole is

Eq 4-4 $$\varkappa = \frac{\gamma}{\rho},$$

where γ is its surface tension.

---

35 Heffron, Michael. *Ramblings of an Old Man: about the scientific method* (self-published via Kindle Direct Publishing, 2020). Chapter 6 (page 44) sloppily referred to that as the Schwarzschild radius.

> Another consequence of Eq 4-3 is that the surface tension of any given black hole divided by the pressure of the aether (P) determines the luminiferous radius (R_L) of that black hole,
>
> **Eq 4-5** $$R_L = \frac{\gamma}{P}.$$

It is important to notice that the surface tension of black holes per density of the aether determines the gravitational parameters of all bodies—from galaxies, down to stars, down to planets, down to moons, down to atoms, down to subatomic particles, and probably beyond. Similarly, the surface tension of any given body per the pressure of the aether determines its luminiferous radius. Future chapters reveal that the surface tension of black holes produces not only gravity and size, but also electricity, magnetism, and light.

Table 2–1 makes it clear that mass is momentum per velocity and energy is momentum times velocity. Table 2–1 makes it equally clear that density is mass per volume and pressure is energy per volume. From the dimension and velocity multipliers of Table 2-1, it is deducible that the gravitational parameter (velocity squared times distance) is surface tension per density (Eq 4-4), and luminiferous radius is surface tension per pressure (Eq 4-5).

Gravity and pressure

Many people like to say a satellite "falls" toward a planet at exactly the same rate the surface of the planet "falls" away from the satellite. That is an accurate description of the constant spacing between concentric rings for a perfectly circular orbit, but it fails to provide a satisfying explanation for the source of the presumed "force" of gravity.

Imagine a satellite in orbit along the outer concentric ring. That satellite experiences a pressure relative to the aether with which it collides. That pressure is similar in concept to the lift developed by an aircraft wing; however, unlike air, aether has zero viscosity–so the pressure experienced by the satellite is frictionless (lacking drag).[36]

Now, imagine that a black hole is like a spherical drain, circumscribed by the inner concentric ring. At any given distance from the drain, the pressure of the inflowing aether pushing a satellite toward the drain (into the black hole) is counterbalanced by the surface tension (pressure of the aether times the luminiferous radius) of the black hole at the satellite's distance from the drain.

---

36 Interested readers should research the role of Euler equations in fluid dynamics, especially the "streamline curvature theorem," while taking note of the many controversies surrounding the source of lift and the validity of Euler's equations.

Every black hole has a pressure gradient (P) defined by

Eq 4-6
$$P = \frac{\gamma}{r},$$

where γ is the surface tension of the black hole, and r is the distance from the surface of the black hole (or radius of orbit). Thus, it is reasonable to view that pressure as the mass flow per second (surface tension) that drains into the black hole at that distance.

The satellite is *weightless* whenever the aether pressure lifting/pushing the satellite away from the drain exactly counterbalances the aether pressure sucking/pushing the satellite toward the drain. That may be very similar to how pressure differences between the bottom and top of a wing enables flight at various altitudes, or buoyancy within water pressure gradients enables a submarine to "hover" underwater, or buoyancy of a hot air balloon allows it to "float" within the air.

Continuous mass flow into black holes causes them to develop pressure gradients equal to their surface tension (mass flow per second) divided by the distance beyond their surface. Satellites develop a pressure equal to their velocity squared times the density of the aether.

Relative to the aether, the pressure of any moving object is always

Eq 4-7
$$P = \rho v^2,$$

where ρ is the density of the aether, and v is the velocity of the object relative to the aether. Note that Eq 3-2 is merely this equation when the velocity is the speed of light.

Just as a hot air balloon floats to an air pressure equal to its own, a satellite floats to a radius where the pressure within the gradient of the black hole equals that of the satellite. Due to the gravitational parameter of the black hole, the velocity of the satellite decreases as it floats farther from the black hole, and increases as it floats closer to the black hole. Because pressure is flowing density, those velocity changes produce corresponding pressure changes.

Relativity

Everything we call constant is actually variable but stable. That includes the density and pressure of the aether, and the resulting speed of light, as well as the gravitational parameter for each black hole.[37] Many of our most important modern "standards" of measurement (such as distance, mass, and time) are based upon the variable but stable speed of light.

That means everything (including us) expands and contracts with the momentum of the aether of which we consist, and that is why those variable "standards" appear to be constant. One very real basis for relativity is that all of our measurement standards for mass and spacetime are actually variable rather than constant.

The remainder of this chapter describes how gravity binds particles together and produces the states of matter. Although the following topics may be interesting and help with visualization of the aether, they are not essential to understanding the important topics of this book.

States of matter

To paraphrase Mister Ray (the author's high school physics teacher), matter has the following four known states: 1) Solids have a definite shape and volume; 2) Liquids have a definite volume, but conform to the shape of their container; 3) Gasses have neither definite shape nor volume and they readily escape from any open or permeable container; and 4) Plasma is a nebulous, electrically conductive state of matter that often results from exposing a gas to high voltage—thereby causing its molecules to separate from each other and glow with a characteristic color.[38]

The next few topics examine matter and its behavior under the influence of aether. Each topic frames well known behavior in terms of how the aether produces that behavior.

Binding forces

Chemists and physicists refer to the various binding forces using various terms, but it is the tremendous mass flow of aether toward the black holes within the nuclei that actually forces ("binds") subatomic particles so tightly together.[39] That

---

37 Table 2–1 indicates that density and pressure are just velocity variations from their shared mass flux (which is the variable but stable momentum per volume). The propagation speed of light is the square root of the variable but stable pressure of the aether divided by its variable but stable density. For any given black hole, its gravitational parameter is equal to its variable but stable surface tension divided by the variable but stable density of the aether.

38 Sir William Crookes called it "radiant matter" (due to its glow) in the late 1800s, and Irving Langmuir named it "plasma" in the 1920s (because it reminded him of the characteristics of blood plasma).

39 The enormous pressure of the aether near a molecule results from the surface tension (mass flow per second) of its various atomic and subatomic black holes relative to the distance from the various surfaces.

gravitational binding keeps the molecules of a solid together despite the kinetic energy of their own internal vibrations trying to spread them apart. For solids, the aethereal binding pressure is so strong that lifting or rotating any portion of the solid lifts or rotates the entire solid. Gaps between "solid" molecules are typically smaller than the molecules, so they strongly resist allowing the molecules to wander around or other solids to move among them.

Gaps between "liquid" molecules are typically about the size of the molecules, which allows them to freely move past each other. Lifting a liquid's open container lifts the entire liquid; however, rotating that container enables gravity to pour the liquid out of the container.

Viscosity is a measure of resistance to fluid flow. Low viscosity liquids pour easily and only weakly resist allowing solids to pass through them; however, other liquids may or may not experience resistance. For example, miscible liquids easily mix, whereas immiscible liquids do not mix. The more viscous a liquid is, the more it behaves like a solid (narrower gaps between molecules). For example, glass may be an extremely viscous liquid.[40]

Both solids and liquids pass fairly easily through even the densest gasses. The concept of miscibility doesn't generally apply to gasses, because the gaps between gas molecules are normally much larger than the molecules themselves.

Stars, lightning, and neon signs are all examples of plasma, which may offer the least resistance of all to lower states of matter; however, that freedom of mobility results from disassembly of the molecules by the extremely high energy.

State transitions

Heating a substance increases its kinetic energy, generally causing it to expand. Eventually the expansion creates enough space (gaps) for formerly "bound" molecules to move more freely. It is the gravity of the black hole(s) within the nucleus that binds electrons to their atoms.[41] It is now clear that mass flow of the external aether toward the nuclei pushes together ("binds") the molecules of the transitioning substance, similar to the way that gravity binds objects to the earth. The kinetic energy of the applied heat is mostly expended to give molecules of the transitioning substance enough momentum (which many view as kinetic energy in this case) to overcome the inflowing aether that keeps them bound.

Because temperature is proportional to kinetic energy, temperature remains fairly constant while a substance is transitioning. That happens because the kinetic energy of the applied heat primarily gives transitioning molecules enough momentum to pass through the gaps rather than allowing the gaps to widen.

---

40 Some scientists argue that glass is an amorphous solid that is not quite liquid.
41 Heffron, Michael. *Ramblings on an Old Man: about the scientific method* (self-published via Kindle Direct Publishing, 2020). Chapter 5 (page 37).

Transition temperatures

A solid approaches its melting point as the temperature increases and the continued expansion pushes the molecules far enough apart to form gaps where the molecules can begin to move past each other. As the gaps between molecules slowly widen, the solid first softens and then begins melting into pools of liquid. For the duration of that state transition, the temperature (kinetic energy) remains fairly constant until all of the formerly solid molecules liquify by attaining that kinetic energy level. For example, ice remains at about 0°C (32°F) while it is melting. During the transition, while the temperature remains constant, the heat applied to the solid produces enough kinetic energy to enable increasingly more molecules to achieve enough momentum to move freely past each other throughout the newly forming liquid.[42]

After its transition from solid to liquid, continuing to heat a liquid increases its kinetic energy (thus causing it to expand as its temperature rises again) until it approaches the required temperature for transition from liquid to gas. Just as occurred for the solid, as the temperature of a liquid rises, the molecules become increasingly mobile–then high momentum particles begin escaping from the surface tension of the liquid as it vaporizes into a gas. For the duration of the transition, the temperature remains constant–with all of the added heat energizing the liquid molecules as they escape to make the transition into gas.

After all of the liquid evaporates into gas, continuing to heat the gas causes the temperature to rise again (due to increasing kinetic energy) until it reaches the required temperature for transition from gas to plasma. Further heating transforms high momentum gas particles into a plasma, breaking atomic bonds, causing the ionized gas to conduct electricity, and causing the plasma to glow the characteristic color(s) of its constituent atoms.

Dimensional cages

The solid state confines molecules within a three-dimensional cage of aether pressure. The liquid state confines molecules within a two-dimensional cage of surface tension. The gaseous state confines molecules within a one-dimensional cage of force. The plasma state disassembles molecules and frees their constituent particles from their cages, thereby releasing them as non-dimensional potential energy.

Notice how the states of matter correspond to the dimensions of the rightmost column of Table 2–1. Notice also that pressure is surface tension (mass flow per second) per distance, force is surface tension times distance, and energy is surface tension times area.

---

42 Interested readers may wish to independently research "enthalpy" and related thermodynamic topics.

As the mass of the aether flows into the atomic and subatomic black holes that comprise the molecules of a substance, it is the surface tension per distance from the black holes that determines the size of the various gaps that enable state transitions to occur.

Vaporous reality

For mass and energy to behave as they do, aether must consist of a large number of small particle-like spheroids, which move unhindered between collisions. The gaps between aether particles are much larger than the particles themselves, which causes zero viscosity (no resistance to fluid flow) on the scale of the gaps between aether particles.

Our thought processes can be significantly impacted by whether we view pressure as energy per volume, force per area, surface tension per distance, or some other kind of fluid flow. **Because aether behaves like a gas, it is important to realize that even what seems most solid in our existence is actually swirls of vapor.** Interested readers may wish to independently research Navier-Stokes and similar equations to better understand multidimensional and/or complex fluid flows.

Limitless energy

At the time of this writing (in 2023), total daily energy consumption by the entire population of the world is about 2e18 joules (J), which is the energy of aether pressure applied to a volume about the size of the smallest specks observable by optical microscopes. The aether within which we live is a seemingly limitless pool of energy, just as Nikola Tesla claimed. It may be easiest to harvest the amount of energy we currently use in the form of plasma, hence a possible motive for why lightning (and the plasma generated by his coils) so fascinated Nikola Tesla.

Topics for further research

Do we have any basis for believing the density and pressure of the aether are constant throughout the universe? If not, what is the impact on the gravitational parameter of distant black holes (electrons, protons, neutrons, planets, moons, stars, and/or galaxies)? If an intensely hot big bang created the universe, what are the implications for the actual age of the universe given that the initial speed of light would have plummeted as the universe cooled? Would that make a good simulation problem for advanced thermodynamics students? How does temperature of the surrounding aether impact particle size? Does that temperature represent an allowable range for the size of an electron, a proton, or a neutron? What about smaller particles, such as quarks or neutrinos? Could aether be a state of matter that fills the expanse from above plasma to beneath solid? In other words, could aether be a fifth state of matter that spans the gap between high pressure plasma and high density solid? Would that enable state transitions from plasma or solid to aether? Other than it being a medium that supports density and pressure, what exactly aether may be is a major topic for further research. Could it be that aether consists of neutrinos (which are about a million times smaller than electrons) or perhaps something even smaller? Is there something smaller still of which neutrinos are made? Are those really "particles" or just even smaller swirls? Consider fractal geometry when answering those last two questions (refer to Appendix B if necessary). Is it helpful or harmful (or a little of both) that international agreement defines the speed of light as a constant, when it is actually a variable determined by pressure and density of the aether? Although it isn't a constant, does treating the ratio of pressure to density as a constant give us more accurate measurements for everything else? Would treating the variable speed of light as a constant give the illusion of relativity? Would treating the variable density and pressure of the aether as a constant give the illusion of relativity? Is everything indeed relative—to the aether?

5 HOW AETHER PRODUCES ELECTROMAGNETISM

Ancient Greeks discovered that using fur to rub amber causes the amber to attract small flecks of various materials. That is due to the phenomenon we now call static electricity. Our modern word "electricity" originates from ηλεκτρον (elektron), the Greek word for amber.

Around 1650, German scientist Otto von Guericke built a crude mechanical electrostatic generator that motivated development of much better electrostatic generators, which stimulated widespread interest in the study of electricity. In 1746, Dutch scientist Pieter van Musschenbroek invented the first primitive capacitor (now called a Leyden jar) for storing high voltage static electricity. Such storage of electricity in a jar came to be known as a "charge" of electricity.[43]

When lightning struck Benjamin Franklin's kite in 1752, electricity flowed down the conductive wet string until it hit a key near the end of the string and jumped from the key into a Leyden jar. From that moment on, it seemed natural to compare electrical "current flow" to water flow, and electrical voltage to water pressure.

In 1820, Danish physicist Hans Christian Ørsted (often alternatively spelled Oersted) accidentally discovered that electricity and magnetism are related. Within months after that discovery, German chemist Johann Schweigger devised the earliest form of electromagnetic galvanometer–which measured small electric currents using the deflection of a magnetic needle in response to electric current passing through a coil of wire.

Soon thereafter, French physicist André-Marie Ampère extensively studied the relationship between electricity and magnetism. His most important contribution came to be called Ampère's law, which states that the mutual

---

43 At that time, the word "charge" simply meant to fill something, most often when filling cargo vehicles or vessels.

attraction/repulsion of two current-carrying wires is proportional to their lengths and to the intensities and directions of their currents.

To honor Ampère's contributions to science, the basic unit of electric current flow is called the ampere (A). **One ampere of electric current flow produces a magnetic tension of 2e–7 N/m between the wires that carry the current.**[44] That tension repels wires that carry current in the same direction, or attracts wires that carry current in opposite directions.

Over the years, the word "charge" came to mean the property of matter that produces the tension when placed in an electromagnetic field. For a single electron or proton, that is an *elementary charge* (*q*) of 1.60218e-19 As (ampere·second).[45] Thus, it is reasonable to view the elementary charge as a way to specify tension (mass flow per second) using easier to measure electromagnetic units.

Electric and magnetic constants

Working together, the electric and magnetic constants define electric and magnetic properties of mass flow resulting from the density and pressure of the aether. The story of how aether produces electromagnetism begins with charge squared per electron mass, which expresses mass as ampere·seconds (charge) squared per kilogram.

Using the Greek letter chi (χ) to represent charge squared per mass, the charge squared per electron mass,

Eq 5-1 $$\chi_E = \frac{q^2}{4\pi\rho\S R_E^3} = (2.24245\text{e-9 A}^2\text{s}^2/\text{kg}),$$

is critically important to both the electric and magnetic constants, where q is the elementary charge, ρ is the density of the aether, and $\S R_E^3$ is the volume of an electron. From Eq 3-3, note that $\rho\S R_E^3$ is the mass of the electron (m_E).

Both the electric and magnetic constants result from the surface tension of the proton relative to the mass of the electron (expressed as square ampere·seconds via charge squared per electron mass).

Just as the density and pressure of the aether produce gravity, mass, and energy–they also produce electromagnetism. All of those attributes of the momentum of aether are inseparable. They are merely different ways to view the aether, many of which are indicated by Table 2–1.

---

44 N/m is newtons of force per meter of wire, which has the same units as surface tension (kg/s²).

45 One ampere is an electrical current flow rate of one Coulomb (C, which is 6.24151e18 elementary charges) per second. The current (mass) flow is what produces a magnetic tension of 2e–7 kg/s² between the wires.

Temporarily foregoing conversion of mass into electromagnetic units, **the electric constant (distributed capacitance of free space) is the density of the aether per surface tension of the proton.** Thus, the electric constant is the reciprocal of the mass of an electron times the gravitational parameter of a proton (which was obscured by conversion into electrical units).

The electric constant (ϵ_0) is

Eq 5-2
$$\epsilon_0 = \frac{\rho \chi_E}{\gamma_P} = \frac{\chi_E}{\varkappa_P},$$

where ρ is the density of the aether, χ_E is charge squared per electron mass (Eq 5–1), and $\gamma_P = PR_P$ is the surface tension of the proton (where P is the pressure of the aether, and R_P is the luminiferous radius of the proton). Note that $\dfrac{\rho}{\gamma_P}$ is the inverse of the gravitational parameter (Eq 4-4) of the proton, and that $\dfrac{\rho}{\gamma_P}$ times χ_E is ϵ_0. In other words, the electric constant (ϵ_0) is charge squared per electron mass (χ_E) per gravitational parameter of the proton ($\varkappa_P$). Appendix C provides an extended mathematical analysis that more directly reveals those dependencies and other interesting equivalences.

Temporarily foregoing conversion of mass into electromagnetic units, **the magnetic constant (distributed inductance of free space) is the surface tension of the proton per pressure of the aether.** Thus, the magnetic constant is the mass of an electron times the luminiferous radius of the proton (which was obscured by conversion into electrical units).

The magnetic constant,

Eq 5-3
$$\mu_0 = \frac{\gamma_P}{P \chi_E} = \frac{R_P}{\chi_E},$$

where $\gamma_P = \rho \varkappa_P$ is the surface tension of the proton, ρ is the density of the aether, $\varkappa_P$ is the gravitational parameter of the proton, P is the pressure of the aether, and χ_E is charge squared per electron mass (Eq 5–1).

Thus, dividing $R_P = \dfrac{\gamma_P}{P}$ (Eq 4-5) by χ_E (Eq 5-1) produces μ_0. In other words, the magnetic constant (μ_0) is the luminiferous radius of the proton (R_P) per charge squared per electron mass (χ_E).
Appendix C provides an extended mathematical analysis that more directly reveals those dependencies and other interesting equivalences.

At this point, it is obvious that the density and pressure of the aether explain mass, energy, surface tension, gravity, electricity, and magnetism. There are many valid ways to view the electric and magnetic constants; however, it is the aethereal view that leads to true comprehension of those constants. Expression as electromagnetic units nearly conceals the fact that **density of the aether produces the surface tension that causes magnetism and pressure of the aether produces the surface tension that causes electricity**.

Electric and magnetic fields

Density of the aether not only explains the magnetic constant, it also explains magnetic fields. Pressure of the aether not only explains the electric constant, it also explains electric fields.

Pressure of the aether mobilizes electrons, thus causing electron flow. Density of the aether inhibits electron flow, thus producing magnetism. Electromagnetic waves (such as light, radio, and x-rays) are oscillations between the pressure (rarefaction, electricity) and density (compression, magnetism) of the aether.

Advanced readers may wish to explore the electric **D** and **E** vector fields,

Eq 5-4 $$\mathbf{D} = \epsilon_0 \mathbf{E},$$

in terms of how electrons become free from the gravitational parameter of the proton, $\epsilon_0 = \dfrac{\rho \chi_E}{\gamma_P} = \dfrac{\chi_E}{\varkappa_P}$, as defined by Eq 5-2.

Similarly, it may be valuable to explore the magnetic **B** and **H** vector fields,

Eq 5-5 $$\mathbf{B} = \mu_0 \mathbf{H},$$

in terms of how the luminiferous radius of the proton captures electrons, $\mu_0 = \dfrac{\gamma_P}{P \chi_E} = \dfrac{R_P}{\chi_E}$, as defined by Eq 5-3.

It is highly advantageous to view things backwards, forwards, from the inside, and from the outside. It is the reconciliation of those various vantage points that leads to true understanding.

Maxwell's equations

Within decades of Ørsted's discovery of electromagnetism, Scottish physicist James Clerk Maxwell developed a coherent theory of electromagnetism by consolidating and extending prior discoveries by Ampère, Faraday, and Gauss.

Advanced readers may wish to examine Maxwell's calculus equations (mentally translating "electric" into aether pressure and "magnetic" into aether density) for a better understanding of complex, multidimensional

---

electromagnetic flows. Also reconcile those flows with how the aether produces electric and magnetic "fields" (Eq 5–1 through 5–5). The aethereal view is that Maxwell's equations deal largely with the convergent and divergent flow of aether resulting from the surface tension of black holes (electrons and protons).

Surfing a light beam

Imaginary perspectives can be very enlightening, but can also be very misleading if mistaken for reality. Instead of imagining yourself riding on a beam of light with Albert Einstein, imagine yourself surfing with Nikola Tesla on the crest of a longitudinal wave of light propagating through the aether. Pressure of the light (electric behavior) compresses the aether into a dense obstructing wavefront (magnetic behavior) that propagates through the frictionless aether until it dissipates back into equilibrium (much like an ocean wave "reaching" a beach).

Electricity and magnetism have such an inseparable relationship because they derive directly from the inseparable relationship between pressure and density of the aether (velocity variations within their shared momentum of the aether). It can be very insightful to perform thought experiments on those relationships, but never mistake thoughts for reality!

Electric pressure and magnetic density

It is completely valid to view the universe as electromagnetic, but far more enlightening to view the universe as a tranquil equilibrium of aethereal momentum that is perturbed by velocity changes. Those manifest as density and pressure changes that we perceive as mass, energy, temperature, surface tension, gravity, electricity, magnetism, and light. It is important for readers to understand that **velocity variations of the momentum of the aether, applied to various dimensions, produce all of the views that we interpret as the physical attributes of the universe.**

Longitudinal electromagnetic waves

The pressure of the aether creates the rarefaction that compresses surrounding aether into high density areas. Density is mass obstructing pressure's path. Density moderates pressure by getting in its way. It is the nature of the aether to constantly seek equilibrium between density and pressure (via conservation of momentum). It is that characteristic that makes electricity and magnetism inseparable–as longitudinal pressure waves (electric) move through the aether, they compress obstructing aether into longitudinal density waves (magnetic).

It helps to visualize particles of the aether much like air molecules randomly bouncing off each other to create the air pressure. When a sound wave propagates, it builds the pressure, which compresses obstructing air molecules, thereby increasing their density. In other words, the pressure of the sound wave produces a dense crest along its leading and trailing edges, encapsulating the sparse pressurized rarefaction between those spherical crests.

Just as Tesla said, light waves in the aether behave like sound waves in the air. Both aether and air molecules naturally tend toward the most uniform distribution possible of density and pressure. That is precisely why sound and light are longitudinal waves that ripple outward from the source and diminish in intensity with distance.

Mass of a particle is its volume times the density of the aether (momentum below equilibrium, a magnetic property). Potential energy of a particle is its volume times the pressure of the aether (momentum above equilibrium, an electric property). The surface tension of a particle results from aether density times its gravitational parameter in electromagnetic equilibrium with aether pressure times its luminiferous radius.

Notice that aether particles, whatever they may be, collide and bounce off each other–but maintain the average density and pressure of the aether–even when filling the volume of a particle to produce mass and energy, or when applied to the surface tension of a body to produce its luminiferous radius, gravitational parameter, electricity, magnetism, and light.

Topics for further research

Is electricity merely low potential plasma that finds it easier to travel within a "conductor" rather than jump the gap into less conductive air? From just the revelations up to this point, it should already be clear to many readers that enormously complex computer simulations can be greatly simplified. For any readers so inclined, be sure to examine Appendix A. The author highly recommends for anyone interested in computer simulations to carefully examine the roles of density, pressure, temperature, and velocity flow (four different views of the same momentum). It is exceptionally clear those are all highly relevant interacting parameters of the aether.

6 LIGHT VERSUS MATTER

Nikola Tesla devoted much of his life to the study of mechanical and electromagnetic energy, frequency, and vibration.[46] Frequency is the number of repetitions per second of a recurring event (such as the vibrations of a tuning fork, the notes of a musical instrument, the color of a light wave, or the whirls of a subatomic particle). Electromagnetic resonance is a magnetic "field" inducing an electric "field" inducing a magnetic field inducing an electric field ... ad infinitum (until something "damps" the oscillations).

Because electromagnetism is mechanical (velocity fluctuations within the shared momentum of density and pressure), electromagnetic resonance is really a mechanical pressure change (electric) inducing a mechanical density change (magnetic) inducing a pressure change inducing a density change... ad infinitum. Resonance occurs when the conditions are exactly right to sustain an oscillating equilibrium of the momentum (which is oscillating density and pressure fluctuations within the aether).

Resonant electronic circuits can be formed using capacitors and inductors. A capacitor is a component that stores electric energy (pressure times volume), while an inductor is a component that temporarily inhibits electric flow by transforming it into magnetism (density, mass per volume).

In a resonant circuit, the capacitor discharges its energy into the inductor, which quickly reflects that energy back into the capacitor. That would be much like a wheelbarrow full of water running into a brick wall. The water briefly sloshes forward, but then sloshes (oscillates) back and forth thereafter. In like manner, the energy oscillates back and forth between a capacitor and an inductor.

Like the bucket of a wheelbarrow contains the water—a capacitor stores aether pressure for a relatively long duration, whereas an inductor only

---

46 Nikola Tesla famously said "If you want to find the secrets of the Universe, think in terms of energy, frequency, and vibration."

transforms pressure into density for a relatively brief period—like the slosh against a brick wall. It is the brevity of inductive storage that causes resonance when the inductor releases its temporary density as pressure that flows back into the capacitor.

The resonant frequency (f_r) of an oscillating electronic circuit is

Eq 6-1
$$f_r = \frac{1}{2\pi\sqrt{LC}},$$

where L is inductance, and C is capacitance. Appendix C provides many other interesting details about inductance and capacitance.

Electromagnetic resonant circuits are widespread. They give us the frequencies for modern communication systems, timing systems, the computers that control our modern conveniences, and much more. Mechanical resonant devices are also widespread. They give us train whistles, tuning forks, musical instruments, and much more. Could the aether itself have a mechanical resonance between the electric constant (distributed capacitance of free space) and the magnetic constant (distributed inductance of free space)?

The frequency (f) of a whirling black hole is

Eq 6-2
$$f = \frac{c}{2\pi R_L},$$

where c is the speed of light, and R_L is the luminiferous radius. Compare that to Eq 6-1, realizing it is equivalent to $f = \dfrac{1}{2\pi\sqrt{\frac{R_L^2}{c^2}}}.$

It has long been known that matter can produce light. Good examples thereof are ancient fires and modern light bulbs. Only recently (July 2021), SciTechDaily reported that scientists studying particle collisions at the Relativistic Heavy Ion Collider at Brookhaven National Laboratory produced "definitive evidence" that light can produce matter.

Because light is oscillations within the aether, and matter is swirls within the aether, it then makes perfect sense that light can transform into matter and matter can transform into light. That is, a swirl can become an oscillation, and an oscillation can become a swirl. Both are periodic disturbances within the aether. Moreover, it makes perfect sense that out-of-phase oscillations/swirls should be able to annihilate each other via phase cancellation (much like theoretical matter/antimatter), leaving behind only whatever substance comprises the aether.

Light as a form of matter

Light is an electromagnetic phenomenon, an oscillation between the density and pressure of the aether. Just like the whirls of aether we call subatomic particles, **the longitudinal oscillations of aether we call light have a mass, constant gravitational parameter, luminiferous radius, surface tension, and pressure.**

Not only does light have the characteristics of matter, but all of those attributes indicate that the longitudinal oscillations of light behave as if they are more massive than the whirls of subatomic particles with which they interact.

Regardless of its wavelength (λ), light has a constant surface tension (γ_λ) of

Eq 6-3 $$\gamma_\lambda = \rho \varkappa_\lambda = PR_\lambda,$$

where ρ is the density of the aether, $\varkappa_\lambda$ is the constant gravitational parameter of light, P is the pressure of the aether, and R_λ is the constant luminiferous radius of light. Compare this to the surface tension of matter (Eq 4-3), and refer to Appendix C for supplemental information, or Appendix C or D for the actual value of those constants.

Light absorption and emission equalizes the pressure when light interacts with electrons or vice versa. There are many other valid views of light and electrons; however, viewing them from the perspective of pressure due to their surface tension is the most insightful.

For any given wavelength (λ) of light, the pressure of the light is

Eq 6-4 $$P_\lambda = \frac{\gamma_\lambda}{\lambda},$$

where $\gamma_\lambda = \lambda P_\lambda = 3.89136e21$ kg/s^2 is the constant surface tension of light.
  Compare that to the pressure of an electron,

Eq 6-5 $$P_E = \frac{\gamma_N}{r_E},$$

where $\gamma_N = P_E r_E$ is the constant surface tension of the nucleus, and r_E is the radius of the electron's orbit (or its distance from the nucleus). Exchange P_E with r_E and it becomes very clear that an electron floats within the pressure gradient of the nucleus. Note the enormous similarities between light and matter demonstrated by these two equations!

An electron emits a wavelength of light that is equal to the constant surface tension of light per the pressure of the electron prior to emitting that light.

An electron emits light of wavelength

Eq 6-6
$$\lambda = \frac{\gamma_\lambda}{P_E} = \frac{\gamma_\lambda r_E}{\gamma_N},$$

where γ_λ is the constant surface tension of light, and P_E is the pressure of the electron (where r_E is the radius "of orbit" for the electron before emitting that light, and γ_N is the surface tension of the nucleus). Those parenthetical terms are "per the pressure of the electron" (inverse of Eq 6-5).

The radius of an electron's orbit after absorbing light is equal to the surface tension of the nucleus per the pressure of the light.

The radius of an electron's orbit after absorbing light is

Eq 6-7
$$r_E = \frac{\gamma_N}{P_\lambda} = \frac{\lambda \gamma_N}{\gamma_\lambda},$$

where γ_N is the surface tension of the nucleus of the atom, and P_λ is the pressure of the light (where λ is the wavelength of the light, and γ_λ is the surface tension of light). Those parenthetical terms are "per the pressure of the light" (inverse of Eq 6-4), which must equal the pressure of the electron (P_E), so

Eq 6-8
$$r_E = \frac{\gamma_N}{P_\lambda} = \frac{\gamma_N}{P_E}.$$

Grand unification

It is the luminiferous aether that unifies all of the forces of nature. Momentum of the aether directly causes density, pressure, mass, energy, temperature, surface tension, gravity, electricity, magnetism, and light. Interested readers may also wish to research the "strong" and "weak" nuclear forces, which are already known to be electromagnetic.[47]

---

47 Ucar, H. Polarity Free Magnetic Repulsion and Magnetic Bound State. Symmetry 2021, 13, 442. https://doi.org/10.3390/sym13030442 may also be of interest when considering the behavior of whirling nuclei.

7 THE TRUE COLORS OF LIGHT

This chapter slightly enhances chapter 4 of the author's prior book because that material is essential for fully understanding the true nature of light and spectral lines.[48]

Australian independent researcher Remus Poradin relentlessly peered deep into the soul of the prism to discover its secrets. Newtonian physics can't explain what he saw![49]

Just over three centuries ago, like three blind men trying to describe an elephant from the different appendages they held–Sir Isaac Newton, Christiaan Huygens, and Johann Wolfgang von Goethe each described different aspects of a prism without comprehending its whole.

Newton believed the prism bent different colors of light at different angles to produce a rainbow. Huygens argued light was a wave, rather than the corpuscular spray of spinning particles claimed by Newton. Goethe noticed prisms only produced colors at the transitions between dark and light.

The View From Newton's Eye

In his book Opticks, Sir Isaac Newton thoroughly described his mistaken belief that refraction caused the well known ability of the prism to produce a rainbow of colors. Figure 7–1 (which is Figure 2 of his book) is part of Newton's explanation of how refraction also produces the virtual image seen through a prism.

---

48 Heffron, Michael. *Ramblings of an Old Man: about the scientific method* (self-published via Kindle Direct Publishing, 2020).

49 Explore his blogspot (https://remusporadin.blogspot.com/) for the details of his many amazing experiments.

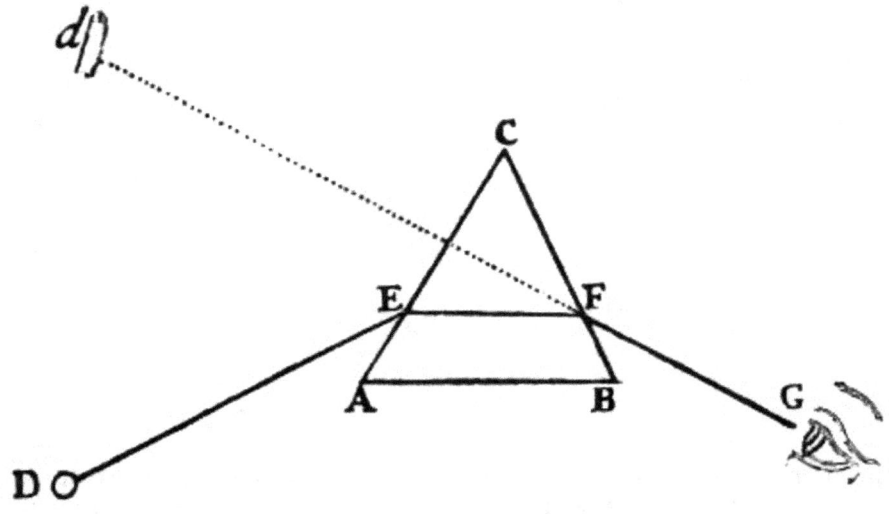

Figure 7–1: This is Figure 2 from Newton's book Opticks.

To paraphrase, Newton said the light from object D travels to point E, where prism ACB refracts it into beam EF, which the air refracts again as it exits the prism at point F, so the observer views the light at point G. Newton explained that double refraction causes the observer to see the virtual image of the object at point d.

Mistakes are okay!!! Science often involves guesswork, and over three centuries ago it was a reasonable initial guess that some kind of refraction might be bending different colors of light at different angles–but remember, **science is always wrong!** Science is a journey, not a destination. It is essential to always look for any anomalies, and explore those anomalies when found. It is also important to view everything from multiple perspectives–top down, bottom up, inside out, outside in–and be sure to occasionally leap completely out of the box!

The View Through Newton's Window

If you can locate a prism, a simple experiment will quickly demonstrate what prisms really do. If you hold a prism with the apex up above your eyes and view a well-lit window surrounded by a dark (or at least dimly lit) wall or door, you can adjust your viewing angle until you observe a blue fringe just *above* the top of the window, a cyan fringe *at* the top of the window, a yellow fringe *at* the bottom of the window, and a red fringe just *below* the bottom of the window. It may take a few minutes of viewing and some keen observation skills to identify those fringes and where they occur. To many eyes, the blue looks more purple or violet, especially against a non-black background (which may mix the blue with dim red). Careful observation and analysis also suggests there may be two slightly different shades of blue.

As illustrated by Figure 7–2, contrary to what modern physics courses teach, the prism separates white light into red, green, and blue panes. It also *appears* to deflect blue toward the apex, red toward the base, and doesn't deflect green at all. Upcoming experiments will confirm all of those facts, plus much more.

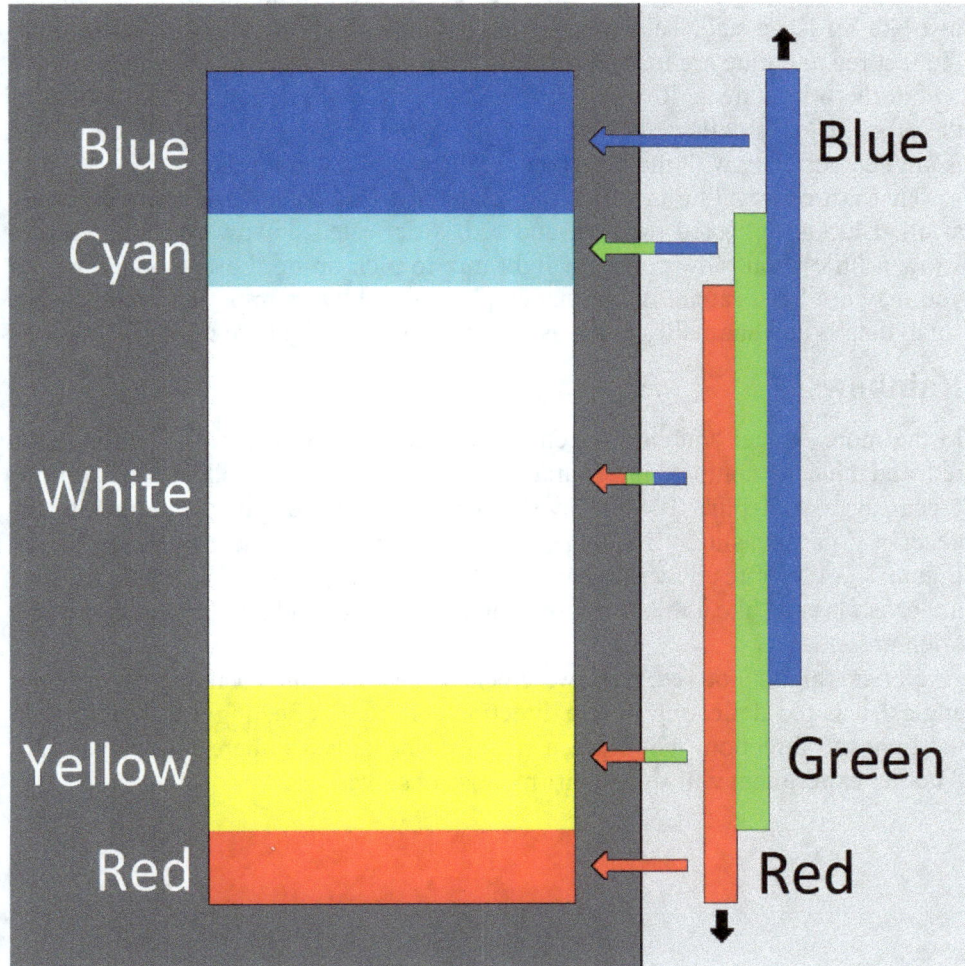

Figure 7–2: Deflection of color panes by a prism.

For any readers who are artists, please note that colors of light don't mix in the same ways as colors of pigments. For example–whereas mixing yellow and blue pigments makes green, mixing yellow and blue light makes white (deducible from Figure 7–2).

That is your introduction to just one of the many things Remus Poradin discovered–an observation that Newtonian physics can't explain. Goethe had it most correct–for, if you place a thin object (such as a pencil) horizontal, fairly close to the prism, between it and the window, you will see a yellow fringe above the pencil, a red fringe along the top, a blue fringe along the bottom, and a cyan fringe beneath.

Your first thought might be–different colors are bending at different angles, just like Sir Isaac said. Take a better look! Think about it. Pay close attention to the fact red and blue are in the areas that were dark, whereas cyan and yellow are within the bright areas. Red is deflected away from green, toward the bottom of the prism. Blue is deflected away from green, toward the top of the prism. Green is aligned perfectly with the light coming through the window.

That last observation is your very subtle clue that green light is not deflected at all. Blue is deflected upward, and red is deflected downward. You have to know a little about how colors of light mix to pick up on those subtle clues, so you may not have caught it–but that emphasizes why it is important to notice the small details and anomalies (such as how the colors of light are mixing).

Rainbows

In the more typical "rainbow" scenario, a much narrower band of light permits red and blue to completely separate from each other to produce red, yellow, green, cyan, and blue. If you have a very steady hand, you can simulate the necessary narrow slit by moving your pencil up toward the top of the prism. Figure 7–3 illustrates how that happens. With a few internal reflections, from an imperfect prism (like a raindrop, for example), orange, indigo, and violet are also often present.

Notice–this is not red, yellow, green, cyan, and blue bending at different angles! It is red deflecting in one direction and blue deflecting a little farther in the opposite direction, while green remains unchanged (where the window is). Another experiment will soon confirm those observations.

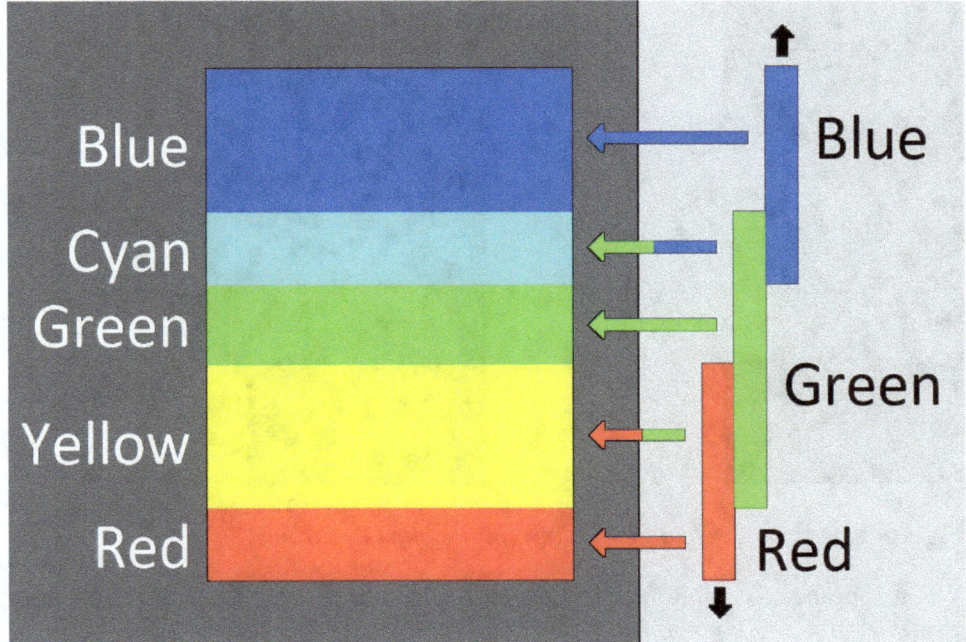

Figure 7–3: Separation of color panes by a prism.

Deflection, not refraction

It may be easier to fully observe color deflection using Figure 7–4(a). Place the prism flat on top of the gray circle, look through the face slanted toward the top of the page, then slowly lift the prism off the page. To keep the circle in view until the metamorphosis occurs, the prism will need to drift toward the top of the page. You can improve the focus by tilting the apex very slightly toward or away from the page. It may also be necessary to move the prism a little closer to or a little farther from the page to get everything clear and crisp. With persistence, you will see color fringes like those depicted in Figure 7–4(b). If you have access to a projector, you can achieve a similar effect by projecting the pattern through the prism onto a close screen or a wall.[50] Note that projected patterns are inverted (which the end of this chapter explains).

---

50 The screen or wall needs to be close enough to keep the colors separated (to prevent the colors from blending or overlapping).

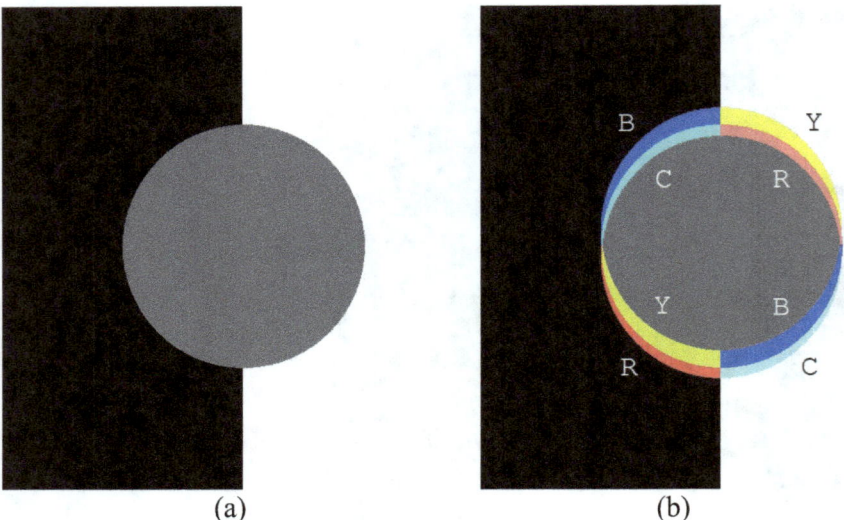

(a)   (b)

Figure 7–4: Gray circle on half black, half white background.

When the apex is pointing up, pay particular attention to the top of the gray circle. Notice the B=blue & C=cyan fringes at the upper-left of the gray circle next to Y=yellow & R=red fringes at the upper-right of the gray circle. Please carefully observe it is the transition from lighter to darker, or darker to lighter, that produces the colors–not refraction!!!

Next, turn your attention to the bottom of the gray circle. Please notice the yellow & red fringes at the lower-left of the gray circle next to blue & cyan fringes at the lower-right of the gray circle. Think about that. Reconcile that observation with the mistaken belief that prisms bend different colors of light at different angles. Thank you Remus for your astute observations and incredibly well crafted yet simple experiments!

Think about why that happened. Realize you see fringes on each side of red and/or blue intensity changes. As subtly demonstrated by the cyan and yellow areas, green intensity changes do not deflect–while red intensity changes deflect toward the base of the prism and blue intensity changes deflect toward the apex of the prism. Look for those effects at the transitions between black, gray, and white.

By rotating Figure 7–4, you will observe it is the transition from darker above to brighter below that shifts blue toward the apex, and the transition from brighter above to darker below that shifts red toward the base. Both red and blue appear to deflect away from the source of the light, toward the darkness–again demonstrating that Goethe's observation was the most correct, although Huygens and Newton were also correct from their own single perspectives. Imagine how the world could have benefited if any of those three would have viewed everything from multiple perspectives. **That is why readers should view** everything from top down, bottom up, inside out, outside in, and occasionally leap completely out of the box.

Black and white lines

Figure 7–5: White line on black background, black line on white background.

Since you now have a better understanding of what the prism is actually doing–Figure 7–5 demonstrates that white light consists of only red, green, and blue–and that the prism deflects red & blue in opposite directions, away from green. Look at the white and black lines through your prism (with its apex facing left or right) and pull the prism back far enough from the page to see the white line completely separates into only red, green, and blue. Similarly, the black line separates into only cyan, magenta, and yellow. The next topic will explain that.

For your convenience, Figure 7–5 shows what colors you will see when the apex of your prism is oriented left or right (R=red, G=green, B=blue, C=cyan, M=magenta, Y=yellow). Notice that blue is slightly wider and deflects slightly farther from the green line than does red. Now, compare that to a typical graph of the light spectrum, which shows just the opposite (red being wider and deflected farther from green). That results from skewing the entire spectrum so the colors will overlap.

Observations in Sunlight

If you are an advocate of Newtonian physics, you may find it extremely unsettling that the white line separates into only red, green, and blue. Similarly, the black line separates only into the absence of red, green, and blue. If you are viewing Figure 7–5 on a color display, you may want to align a black line on white paper with a white line on black paper–and take them out into the sunlight to reassure yourself you are not simply seeing the RGB pixels of your display. If you are already viewing this book on monochrome e-paper or real paper, just take it out into the sunlight to ensure you aren't seeing the effects of artificial lighting.

Carefully observe how the sunlight falling on the thin white line actually separates into thin red, green, and blue lines. Now, observe directly under the green line on the black background is a magenta line on the white background. Magenta = blue plus red. Green is still absent from exactly where the black line was, whereas red and blue deflected in opposite directions to fill in the place that was a black line. **That green is absent only from the black line (where it never was) is how you know green doesn't deflect at all.** Directly above the black line, only an undeflected green line remains where all colors (the white line) once were.

Similarly, observe that directly under the red line on the black background is a cyan line on the white background. Cyan = blue plus green. Red is still absent from where the black line was, but the entire red color pane is now shifted toward the base of the prism.

Finally, observe directly under the blue line on the black background is a yellow line on the white background. Yellow = green plus red. Blue is still absent from where the black line was, but the entire blue color pane is now shifted toward the apex of the prism.

That is exactly how those very same colors of light mix on a color display, which is why you'll want to take lines drawn onto paper out into the sunlight to convince yourself there is no possibility you are seeing anything artificial.

Think about what you just observed. Red deflects exactly the same as cyan (absence of red). Although deflected farther and wider, blue deflects exactly the same as yellow (absence of blue). Green doesn't deflect at all. If you observe carefully, blue deflects in one direction and red deflects in the opposite direction, leaving only green where the white line once was.

As you think carefully about those observations, you will realize blue and red light panes deflect in opposite directions from the stationary green light pane. You are seeing only three colors (two of which are deflected in opposite directions from the third), *not* a rainbow of colors all refracting at different angles. **Note that the behavior of the white and black lines is critically important for understanding absorption and emission spectra in the next chapter.**

Imaginary Colors

If you have ever researched the cones of human eyes, you would know the typical person has specific cones that respond to the intensity of red, green, and blue light. All of the *other* colors we "see" are actually figments of our imagination!

Our minds "see" white when the cones in our eyes simultaneously detect equal intensity red, green, and blue light–striking adjacent rods and cones on our retinas. Similarly, our minds "see" yellow when the light is red and green. When the light is green and blue, our minds "see" cyan. Magenta is what our minds "see" when the light is red and blue.

Think about that long and hard–only red, green, and blue are real colors of light. All other colors are fabrications of our minds. Purple/violet results from dim red and blue. Orange results from red and dim green. Brown results from dark red and dark green.

You might well ask, if only red, green, and blue are real–and all other colors are figments of our imagination, how do we know there aren't other colors we can't see in between those colors? Here's how, another experiment...

Attach various colors of paper to form a thin line on a black sheet of paper– then take that out into the sunlight, view it through a prism, and notice how the colors separate, as depicted by the letters of Figure 7–6. R=red appears deflected toward the base, B=blue appears deflected toward the apex, and G=green remains undeflected.

Figure 7-6. Colors separating into red, green, and blue.

You will also find that O=orange separates into red and dark green, rather than remaining intact and deflecting partway between red and green. Y=yellow separates into red and green, rather than remaining intact and deflecting halfway between red and green. W=white separates into red, green, and blue. C=cyan separates into green and blue, rather than remaining intact and deflecting halfway between green and blue. M=magenta separates into red and blue, rather than remaining intact and deflecting past blue.

Verify everything!

If you're concerned that all colors are digital now–use a petal from a natural yellow flower, cut it into a thin strip, and put it onto a black background. It still separates into red and green. Yellow is your imagination–your mind trying to make sense of incoming red and green light simultaneously hitting adjacent cones of your retina.

By now you should be getting the point–we were all misinformed about what prisms actually do. Yet, despite being totally wrong about a wide spectrum of colors, good engineers have managed to build spectrometers and many other amazing instruments using prisms.

After reading many of the critiques of the incredible experiments performed by Remus Poradin, it becomes clear that most scientists lack the courage to challenge the majority opinion regardless of incredibly clear and convincing evidence the majority is wrong. That so many scientists lack the courage to challenge the majority opinion, dear readers, is precisely why science deludes people into thinking they know everything there is to know–causing them to stagnate in dark ages that should have been long past.[51]

It is the mistaken belief we know everything there is to know that keeps us from looking further to find greater truths. It is easy to imagine a certain "fact" is true when nearly everyone seems to accept it as fact. That deludes us into thinking we "know" things that aren't necessarily true. Readers should verify everything, and always look beyond!

Inverted Spectrum

As the colors begin overlapping and blending a short distance after projection from a prism, the three real colors of Figure 7–3 morph into the familiar colors of a rainbow–red, orange, yellow, green, blue, and violet (ROYGBV). When you look through a prism at a distance that overlaps and blends the colors, you will see a reversed spectrum (VBGYOR). Figures 7–7 and 7–8 illustrate why that peculiar phenomenon occurs.

To ensure clarity about what is actually happening, the following figures use the overlapping, but not blending, RYGCB (red, yellow, green, cyan, blue) color panes of Figure 7–3.

---

51 Heffron, Michael. *Ramblings of an Old Man: about the scientific method* (self-published via Kindle Direct Publishing, 2020). The preface asserts that science stagnates whenever scientists think they know all there is to know.

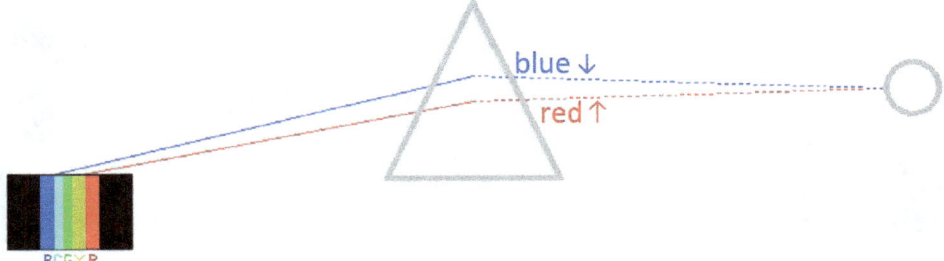

Figure 7–7: White line on black background, viewed through prism.

A prism "lifts" the image that is on the plane of its base. You can demonstrate that by placing the prism next to a pattern, and observe that the prism lifts the pattern into view along the plane of its base.

Although blue appears to the eye to be deflected toward the apex, it is actually deflected toward the base. Similarly, although red appears to the eye to be deflected toward the base, it is actually deflected toward the apex. The deflection we see is the opposite of reality.

Figure 7–7 depicts how an eye, looking through the prism at a white line, appears to see blue above the top of the white line and red below the bottom of the line (thus, producing a BCGYR pattern).

Since the prism deflects blue downward toward the base, the eye must look upward toward the apex to see it. Similarly, since the prism deflects red upward toward the apex, the eye must look downward toward the base to see it. That causes the eye to see the colors in reverse order (BCGYR) from what is projected onto a white wall or screen (next figure).

Figure 7–8 shows that same white line projected through a prism, producing a RYGCB pattern on the wall. That will all make sense when you compare the angles of the colored beams as they exit the prism in Figures 7–7 and 7–8, and realize Figure 7-7 views the deflected light from a single point (the viewer's pupil) whereas Figure 7-8 allows the deflected light to spread across many points on the wall.

projected pattern

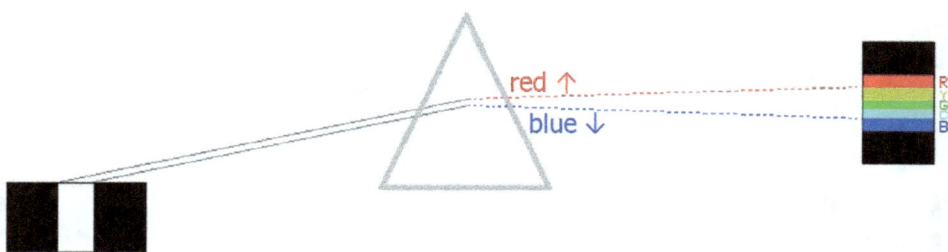

Figure 7–8: White line on black background, projected through prism.

Comparing Figures 7–7 and 7–8 should clarify why those colors appear reversed–depending on whether you are projecting or viewing that image through a prism. It is amazing how many arguments occur over that simple change of perspective. People seem to have enormous difficulty in reconciling a single phenomenon observed from different perspectives–but reconciling those different views is essential for all who seek the truth.

Strive to view everything from the top down, bottom up, inside out, and outside in. You may be amazed by what you discover!

8 SPECTRAL LINES

This chapter refactors chapter 9 of the author's prior book to account for the role played by density and pressure of the aether.[52]

Like various colors of neon lights, gasses such as hydrogen can also be sealed within glass tubes that have electrodes at both ends. When the electrodes are connected to low-current high-voltage electricity, each gas forms a plasma that glows a characteristic color.

Spectroscopy beams the light emitted by glowing plasma through a slit and into a prism to produce spectral lines. Keep in mind that the previous chapter exposed how easy it is for colors to overlap rather than cleanly separate into red, green, and blue lines.

Colors of light are typically measured as wavelength, which is the distance between two successive crests or troughs of the light wave. The simplest method for measuring the wavelength of light involves a diffraction grating and a ruler. Interested readers should independently research that topic. Unfortunately, with diffraction gratings (just as with prisms) it is much easier to overlap colors than it is to cleanly separate them.

Scientists quickly realized each element has a unique set of spectral lines, whose observed wavelengths can positively identify what elements are present. Those scientists also quickly realized spectral lines come in two varieties–black bands where a cold gas absorbs light, and luminous bands where a glowing gas directly emits light. Those bands should evoke thoughts of the black and white lines of Figure 7–5.

Figure 8–1 depicts the visible light portion of the "Balmer m=2 series" of the emission spectrum for hydrogen. In theory, stars begin as huge balls of hydrogen–so the ratio of hydrogen to helium should be a good indicator of the

---

52 Heffron, Michael. *Ramblings of an Old Man: about the scientific method* (self-published via Kindle Direct Publishing, 2020).

age of a star.[53] Because of that, spectroscopy is a favorite tool astronomers use to estimate the ratio of those elements, and to identify which other elements may be present in stars. Due to the importance of spectroscopy to astronomy, a great many prominent scientists theorized about the meaning of those spectral lines, and from that work the theory of quantum mechanics was born.

Figure 8–1: Balmer m=2 series emission spectrum of hydrogen atom.

In 1885, Swiss mathematician Johann Balmer developed a formula to predict the wavelengths of the spectral lines emitted by glowing hydrogen. Other researchers (Lyman, Paschen, Brackett, Pfund, Humphreys, and nameless others) contributed their own series, and the chart began filling with colors.

Balmer discovered the wavelength (λ) of each spectral line of hydrogen could be found by an equation equivalent to

Eq 8-1 $$\lambda_{m<n} = \lambda_{Balmer} \left(\frac{n^2}{n^2 - m^2} \right),$$

where λ_{Balmer} is 3.64507e-7 m, and m is an integer that is always less than the integer n. It is important to note that $\lambda_{0<1} = \lambda_{Balmer}$ is exactly eight times the wavelength (λ_{Bohr} = 4.55634e-8 m) of an electron in orbit at the Bohr radius (a_0 = 5.29177e–11 m) of a hydrogen atom.

   After Balmer discovered the m=2 series of spectral lines, other researchers subsequently discovered the following additional series: Lyman series, m=1; Paschen series, m=3; Brackett series, m=4; Pfund series, m=5; Humphreys series, m=6; plus many other unnamed series, where m>6.

---

53 Readers should be skeptical of that theory, since we now know that stars are black holes that expose the aether. Could stars be the opposite of fusion? That is, could stars be mere condensation from the aether into the cold of space, forming a glowing plasma of subatomic particles that produce primarily hydrogen and helium atoms as the condensates cool? Could the ratio of helium to hydrogen indicate condensation temperature difference rather than the age of the star?

Chapter 7 detailed numerous experiments that clearly demonstrate visible light consists only of the colors red, green, and blue. Other colors appear only when the prism projects the spectral lines so far away that the colors overlap and intermix.

Readers should already suspect something may be amiss with spectroscopy, since many of those spectral colors result from overlapping red, green, and blue. For example, notice how the H–β (cyan) line of the hydrogen spectrum (in Figure 8–1) is nestled between H–γ (blue) and H–α (red). Green is completely absent, and yet cyan (which consists of blue and green) is present. In other words, H–β is a real spectral line for a color that is a figment of our imagination—but there is no spectral line for the real color green, which is necessary for the illusion of cyan!

Since quantum mechanics seeks to explain those same colorful wavelengths, something may be amiss there too! Despite their mistakes, spectroscopy and quantum mechanics seek to explain very real phenomena, so those are areas where tremendous future improvement may occur.

It doesn't matter that mistakes were made! Human knowledge has improved dramatically as a result of all we learned pursuing the wrong paths down which those mistakes led us. **The wonderful thing about true science is that it continually searches for truth, and sometimes finds it!**

Orbital energy

A satellite in orbit is "weightless." Consider that an extremely small force can move a massive weightless object and thereby reveal that orbital energy has radically different characteristics than energy here on earth.[54] Because the gravitational parameter determines velocity (which determines energy), it takes very little effort to accomplish enormous changes to orbital energy.

Now, pause to ponder the discussion (from Chapter 2) about how a beach ball can ride the illusory waves. Consider how easily a light wave might push an electron into a higher or lower orbit to equalize the pressure of the light and the electron, thereby obeying Kepler's Third Law (the constraints of the gravitational parameter).

Photoelectric Effect

Light energy is so precisely known due to the photoelectric effect. When light strikes a curved metal electrode, it ejects the lowest energy electrons. A "stopping voltage" easily suppresses the flow of those electrons and thereby enables precise calculation of their energy.

---

54 Refer to https://www.nasa.gov/audience/forstudents/5-8/features/nasa-knows/ what-is-microgravity-58.html for a description of how astronauts can move massive equipment using only their fingertips.

Velocity squared is a dominating factor within the subatomic realm. It determines pressure (pv^2) and energy (mv^2) directly (because pressure times volume is energy). Thus, the oscillation of a wave of light could easily carry an electron to a different level within the pressure gradient of any given nucleus–and thereby have tremendous influence on the energy of the electron.

The pressures of light (Eq 6-4) and electrons (Eq 6-5) are equal to

Eq 8-2 $$P_{m<n} = P_{Balmer} \left(\frac{n^2 - m^2}{n^2} \right),$$

where $P_{Balmer} = 1.06757e28$ kg/m·s^2 is the pressure of the electron and the light when they interact at the Balmer wavelength of the hydrogen atom, and m is an integer that is always less than the integer n.

It is customary to determine the energy of an electron via the frequency of light times the Planck constant; however, it is equally valid to determine the energy of an electron from its volume times the pressure of light.

Using the Planck constant (h), the energy of an electron is $E = hf$, where f is the frequency of an interacting light wave; however, it is equally valid to find the energy of the electron (E_E) using

Eq 8-3 $$E_E = \left(\frac{\gamma_\lambda}{\lambda} \right) \left(\S R_E^3 \right) = P_\lambda \left(\S R_E^3 \right),$$

where γ_λ is the surface tension of light, λ is the wavelength of the light, R_E is the luminiferous radius of the electron, and P_λ is the pressure of the light (Eq 6-4). That is, the energy of the electron is equal to the pressure of the light times the volume of the electron. Is that a clue to what actually happens when the aether within a wave of light interacts with electrons? What does the invariant volume of the electron imply? For readers who may be interested, Appendix C derives the Planck constant from Eq 8-3.

Quantum levels

It is arguable whether electrons actually orbit the nuclei of atoms, merely exist around those nuclei as bubbles, probability clouds, swirls, vortices, whirls, whirlpools, or something else entirely. Without endorsing any of those theories, this topic explains spectral lines in terms of the pressure of the light, which must equal the pressure of the electron during their interaction.

Table 8–1: Wavelength (λ, in nanometers, nm) and pressure (P) for quantum levels (n) from 12 through 24 of the m=8 series spectrum of the hydrogen atom, with associated RGB levels.

m	*n*	*λ* (nm)	*P* (kg/m · s²)	Red	Green	Blue
8	12	656.112	5.93095e27	255		
8	13	586.682	6.63284e27	255	233	
8	14	541.237	7.18977e27	133	255	
8	15	509.403	7.63907e27		255	23
8	16	486.009	8.00681e27		239	255
8	17	468.188	8.31154e27		160	255
8	18	454.231	8.56693e27		92	255
8	19	443.053	8.78309e27		27	255
8	20	433.936	8.96762e27	40		255
8	21	426.386	9.12639e27	80		255
8	22	420.051	9.26403e27	106		255
8	23	414.674	9.38416e27	118		237
8	24	410.070	9.48953e27	126		219
Balmer		364.507	1.06757e28	($\lambda_{Balmer} = 8\lambda_{Bohr}$)		
Bohr		45.5634	8.54055e28	($P_{Bohr} = 8P_{Balmer}$)		

Table 8–1 depicts the wavelength (Eq 8-1) and pressure (Eq 8-2) associated with theoretical quantum levels *m* and *n*, including the "24-bit" RGB values generated for those wavelengths on a computer display.[55] The pressure of the electron/light times the wavelength of the light is equal to the constant surface tension of the light. It is convenient that *m*=8 maps so well to $\lambda_{Balmer} = 8\lambda_{Bohr}$. An interesting topic for further research might be to determine why visible light only occurs as red, green, and blue.

Notice that the blue band in Table 8–1 is at least twice as wide as red, and deflected just a little farther from green, just as in the prism experiments of Chapter 7. It should already be clear that the wavelength of the light (rather than its quantum level) determines its energy (pressure of the light times the volume of the electron).

Balmer's equation fairly accurately predicts spectral lines via *m* and *n* values– but the fact there are only three actual colors of light indicates that has nothing to do with theoretical quantum levels. That is, **the theoretical quantum levels that correspond to imaginary colors can only be predicting the mixing ratio rather than real colors of light.** Readers who compare the wavelengths generated by various values of *m* and *n* will quickly realize it is the *ratio* of

---

55 Web site https://academo.org/demos/wavelength-to-colour-relationship/ has a wonderful wavelength to 24-bit RGB color converter.

n to m that actually determines the perceived wavelength, without regard to the actual values of m and n. For example, $\frac{n}{m} = 1.500$ corresponds to 656.112 nm red light waves regardless of the values of n and m.

The Balmer equation (Eq 8-1) is equivalent to

Eq 8-4 $$\lambda_{m<n} = \lambda_{Balmer} \left(\frac{n^2}{m^2}\right) \div \left(\frac{n^2}{m^2} - 1\right),$$

which explains why $\frac{n}{m}$ corresponds to any given wavelength regardless of the values of n and m.

For $m=8$, distinctive color mixing so close to display colors was a major revelation. In Table 8-1, observe the following visible color ratios:

$\frac{n}{m} = \frac{3}{2} = \frac{12}{8} = 1.500$ corresponds to 656.112 nm red light waves,

$\frac{n}{m} = \frac{15}{8} = 1.875$ corresponds to 509.403 nm green light waves,

$\frac{n}{m} = \frac{19}{8} = 2.375$ corresponds to 443.053 nm blue light waves, and

$\frac{n}{m} = \frac{5}{2} = \frac{20}{8} = 2.500$ corresponds to 433.936 nm blue light waves.

For figments of your imagination, observe the following color mixture ratios:
$\frac{n}{m} = \frac{13}{8} = 1.625$ corresponds to alternating red and green light waves (the illusion of yellow), $\frac{n}{m} = \frac{4}{2} = \frac{16}{8} = 2.000$ corresponds to alternating green and blue light waves (the illusion of cyan), and $\frac{n}{m} = \frac{6}{2} = \frac{24}{8} = 3.000$ corresponds to alternating red and blue light waves (dim magenta, the illusion of violet).

All ratios of n and m represent how red, green, and blue light waves mix to produce the illusion of a full spectrum of colors. The fact there are only three actual visible colors is what gives the illusion of quantum levels. For all three colors of visible light, plus infrared and ultraviolet, it is pressure (ripples in the aether) that bumps electrons out of orbit, thereby causing the photoelectric effect. When you carefully examine yellow ($n=13$) and cyan ($n=16$) from Table 8–1, the mixing ratios are especially obvious.

There is no need to abandon Balmer, Planck, Rydberg, nor any other useful equation–just take a fresh look and use those equations with greater discernment and understanding of how the aether explains everything.

Topics for further research

Are there two different wavelengths of blue light, or merely different mixing ratios that produce nearly the same wavelength? How many wavelengths are there for infrared and ultraviolet light?

9 EXPERIMENTERS BEWARE!

Chapter 3 mentions the Michelson-Morley experiment, which should be taught as ingenious engineering as well as an embarrassing example of a false negative conclusion. Failing to find a needle in a haystack falls far short of proof that no needle exists!

False positives are also possible, such as mistaking a sharp blade of hay for a needle. That is why Chapter 7 performed experiments to confirm in many different ways that yellow really is an imaginary color composed of a mixture of red and green light. Avoidance of false positives was also the goal that prompted this book to examine the role of the aether as it applies to all of the major "laws" of physics.

This book took the same informal approach to the scientific method as its predecessor.[56] Like that book, it often used the measurements and observations of others, and asked questions about things there is a greater need to understand. The next topic is a very brief excerpt from that book.

The scientific method

Science is the study of the structure and behavior of things. Over the years, science has gotten many things wrong. That's okay! Remember, the nature of science is to study and learn.

Since at least the 17th century, the "modern" scientific method has attempted to formalize the process, reduce the number of mistakes, and leave good documentation to aid the next generation of scientists.

---

56 Heffron, Michael. *Ramblings of an Old Man: about the scientific method* (self-published via Kindle Direct Publishing, 2020).

Depending on who you ask, the modern scientific method consists of something like the following steps:

1) Make observations and/or measurements.

2) Identify something you want to better understand.

3) Form a hypothesis (an educated guess) about what might cause the observed phenomena you wish to understand.

4) Perform experiments to enable you to accept or reject your hypothesis.

5) Document everything!

Some people will argue there are more steps, but those five are the *essential* steps. No matter how many steps you take, repeated application of those steps produces scientific knowledge—but mistakes can be made—you might accept a hypothesis that is actually false, or you might reject a hypothesis that is actually true. In addition, there may be flaws in your data, observations, or analysis.

Experiments gone wrong

The next few topics describe experiments that reached incorrect conclusions, yet each was documented well enough to provide useful information about where each experiment went wrong.

Double slit experiment

As many experimenters are aware, the double-slit experiment produces the seemingly ambiguous result that light appears to behave like both particles and waves. That happens because light *consists of* particles (of aether) displaced *by* longitudinal waves (of aether).

In our reality, light acts massless because longitudinal light waves *consist of net* massless oscillations within the aether. Light acts like particles because light's longitudinal oscillations within the aether **displace** aether particles. Those aether particles have the energy to displace electrons to produce the photoelectric effect. Light behaves as predicted by the theory of relativity because it *is* relative–in that the speed of light is *not* constant. Light behaves like matter because it has the ability to push electrons as though they were beach balls riding in the troughs of ocean waves.

Just as longitudinal sound waves propagate in air at an average speed (which is not a limit), so longitudinal light waves propagate within the aether at an average speed (which is not a limit). "The Original Double Slit Experiment" video by Veritasium (on YouTube) may bring many topics together for interested readers.

Doppler effect

The whistle of a passing train sounds like a higher frequency as it approaches, and like a lower frequency as it departs. That phenomenon is known as the Doppler effect–named in honor of Austrian physicist Christian Doppler, who first described the phenomenon in 1842.

It occurs because the approaching train, moving in the same direction as the sound waves, compresses the waves closer together–thus causing the listener to hear a higher frequency. As the train departs, moving away from the listener, it spreads the waves farther apart–thus causing the listener to hear a lower frequency.

Discovery of Doppler shifts led to the theory that light should experience similar Doppler shifts as it travels through the aether. Many experiments attempted to prove or disprove that theory. Most famous of them all was the Michelson-Morley experiment, which used a unique interferometer, invented by Michelson, to determine the speed of light was not impacted (there was no Doppler shift) by traveling through the aether.

Imagine trying to perform that same experiment on sound waves inside a boxcar of the train, and it will become clear the experiment failed to detect the aether because the earth, their apparatus, and the light it generated were all composed of the very aether whose Doppler effect they were attempting to measure. It seems that Nikola Tesla was the only scientist of his era who understood that. His peers soon stagnated, thinking they knew all there was to know–because the subsequent theory of relativity did such a good job of explaining everything.

Display colors

Color display manufacturers tend to prefer wavelengths of 650 nanometers (nm) for red, 510 nm for green, and 440 nm for blue. Thus, red is 140 nm longer than green and blue is 70 nm shorter.

Consider the burst of pressure that creates the leading and trailing edges of 510 nm green light. If the source of that green light travels 70 nm away from an observer during its propagation, that will add 140 nm to the wavelength the observer perceives at its trailing edge. Similarly, if the light source travels toward the observer, that will subtract 70 nm from the wavelength the observer perceives.

Might those facts imply that 510 nm green light whose source is traveling away from the observer could be Doppler shifted into 650 nm red light? If the source of green light is traveling toward the observer, could it be Doppler shifted into 440 nm blue light?

In other words, is green the one true color of visible light—while red and blue light are merely Doppler shifts? Consider the ability of a prism to "lift" the image that is on the plane of its base. Does a prism tilt red light toward the apex because the trailing edge of its wavelength is farther away from the base (thus making it appear to emerge from the darkness)? Could the prism similarly tilt blue light toward the base because the trailing edge of its wavelength is closer to the base (thus making it appear to depart into the darkness)? Despite his abrasive personality, Goethe had it so right!

Final topics for further research

Is it feasible to build a whirling reflective apparatus to test whether green light can be Doppler shifted into red and blue? Is there a simpler way to test whether Doppler shifted green light produces red and blue? During the interaction and the 70 nm displacement, is there a correlation between the velocity of an electron and/or the number of orbits of the electron as the 510 nm wavelength of green light passes by?

A Frink Programming

Frink is an object-oriented programming language that supports many advanced capabilities. Visit frinklang.org to learn all about the Frink programming language.

This book uses Frink primarily to track units of measure during its calculations to help ensure results are correct. Frink also facilitates conversions to alternative units. For example, it can convert kilograms to pounds, meters to feet, etc.

For the various Frink classes in this appendix, some statements are too long and wrap around to the next line. To avoid confusion over line wrapping, this appendix mimics retro green and white printer paper.

Aether.frink

```
// Defines and provides properties of Aether

// Define units that are currently unknown to Frink
m^-2 s^-1 kg ||| mass_flux

class Aether
{
    // Shape of a particle (sphere assumed)
    class var shapeFactor = 4 * pi / 3

    // Defined by NIST (CODATA)
    class var speedOfLight = 2.99792458000e8 m/s
    class var massOfProton = 1.67262192369e-27 kg
```

```frink
    // Classical ELECTRON radius = luminiferous radius of
PROTON
    class var radiusOfProton = 2.8179403262e-15 m

    // Density, mass flux, and pressure of the aether
    class var density = massOfProton / (shapeFactor *
radiusOfProton^3)
    class var massFlux = density * speedOfLight
    // pressure = massFlux * speedOfLight
    class var pressure = Aether.density * speedOfLight^2

    // Ratios (speed of light squared, and inverse thereof)
    class densityOverPressure[] := density / pressure
    class pressureOverDensity[] := pressure / density
}
```

Aether.frink determines the density of the aether from the mass of the proton divided by its volume[57] and determines the pressure of the aether from its density times the speed of light squared (per Eq 3–2). The following test program demonstrates how to use the Aether class.

AetherTest.frink

```frink
// Tests the Aether class

use Aether.frink

setPrecision[12]

println["Pressure of aether   = " + Aether.pressure]
println["Mass flux of aether  = " + Aether.massFlux]
println["Density of aether    = " + Aether.density]
println["\nSpeed of light three ways...\n"]
println["Pressure / mass flux = " + Aether.pressure /
Aether.massFlux]
println["Mass flux / density  = " + Aether.massFlux /
Aether.density]
println["√(Pressure / density)  = " +
Aether.pressureOverDensity[]^(1/2)]
println["Its inverse = " +
Aether.densityOverPressure[]^(1/2)]
```

---

57 Heffron, Michael. *Ramblings of an Old Man: about the scientific method* (self-published via Kindle Direct Publishing, 2020). Chapter 7 (pages 49-50).

OUTPUT

Pressure of aether = 1.60381845020e+33 m^-1 s^-2 kg
(pressure)
Mass flux of aether = 5.34976250202e+24 m^-2 s^-1 kg
(mass_flux)
Density of aether   = 1.78448868851e+16 m^-3 kg
(mass_density)

Speed of light three ways...

Pressure / mass flux = 2.99792458001e+8 m s^-1 (velocity)
Mass flux / density = 2.99792458e+8 m s^-1 (velocity)
√(Pressure / density) = 2.99792458000**364**e+8 m s^-1
(velocity)
Its inverse = 3.335640951976**0966**e-9 m^-1 s (unknown unit
type)

Note the inaccuracy of the very precise last three digits of the speed of light and last five digits of its inverse! That inaccuracy occurs because Aether.frink only specifies the mass of the proton to twelve digits, thereby causing a corresponding inaccuracy in the density and pressure of the aether at or beyond twelve digits. **Never mistake precision for accuracy!**

BlackHole.frink

```
// Provides properties of a Black Hole
use Aether.frink

// Define units that are currently unknown to Frink
m^3 s^-2 ||| gravitational_parameter
s^2 kg^-1 A^2 ||| charge_squared_per_mass

class BlackHole
{
   class var elementaryCharge = 1.602176634e-19 C
   var howSet
   var myName
   var myRadius = 0

   // Angular velocity
   class angularVelocity[] := Aether.speedOfLight / myRadius
```

```
// Charge squared per black hole mass
class chiFactor[] := elementaryCharge^2 / (4 * pi *
mass[])

// E = mc^2 energy of black hole
class energy[] := Aether.pressure * Aether.shapeFactor *
myRadius^3

// Frequency (Hz) of black hole
class frequency[] := Aether.speedOfLight / (2 * pi *
myRadius)

// Gravitational parameter of black hole (Kepler's Third
Law)
class gravitationalParameter[] := myRadius *
Aether.pressureOverDensity[]

// Acceleration of gravity at specified radius
class gravity[atRadius is length] :=
gravitationalParameter[] / atRadius^2

// Identify how radius was set
class howSet[] := howSet

// Mass of black hole
class mass[] := Aether.density * Aether.shapeFactor *
myRadius^3

// Name of black hole
class name[] := myName

// Luminiferous radius of black hole (which is not
2GM/c^2)
class radius[] := myRadius

// Surface tension of black hole
class surfaceTension[] := Aether.pressure * myRadius

// Volume of black hole
class volume[] := 4 * pi * myRadius^3 / 3

// Force of vortex that rotates black hole
```

```
class vortexForce[] := Aether.pressure * pi * myRadius^2

// Constructor for name and length or mass
new[objName is string, objArg] :=
{
    myName = objName

    if objArg conforms length
    {
        myRadius = objArg
        howSet = "radius directly specified"
    } else
    {
        if objArg conforms mass
        {
            myRadius = (objArg / (Aether.density *
Aether.shapeFactor))^(1/3)
            howSet = "radius set using mass"
        } else
        {
            if objArg conforms gravitational_parameter
            {
                myRadius = objArg /
Aether.pressureOverDensity[]
                howSet = "radius set using gravitational
parameter"
            } else
            {
                if objArg conforms frequency
                {
                    myRadius = Aether.speedOfLight / (2 * pi
* objArg)
                    howSet = "radius set using frequency"
                } else
                {
                    howSet = "unable to set radius"
                }
            }
        }
    }
}
```

BlackHole.frink demonstrates how the luminiferous radius of a black hole determines the following characteristics: mass (Eq 3–3), energy (Eq 3–4), gravitational parameter (Eq 4–2), surface tension (Eq 4–3), volume (of sphere), etc. The following test program demonstrates how to use the BlackHole class.

BlackHoleTest.frink

```
// Tests the BlackHole class

use Aether.frink
use BlackHole.frink

setPrecision[8]

// Create BlackHole objects
neutrino =   BlackHole.new["Neutrino", 2.14e-37 kg]
electron =   BlackHole.new["Electron", 9.1093837015e-31 kg]
protonM =    BlackHole.new["ProtonM",  1.67262192369e-27 kg]
protonR =    BlackHole.new["ProtonR",  2.8179403227e-15 m]
neutron =    BlackHole.new["Neutron",  1.67492749804e-27 kg]
light =      BlackHole.new["Light",    2.18065888937262e5 m^3
s^-2]
bohr =       BlackHole.new["Bohr",     5.29177e-11 m]
earth =      BlackHole.new["Earth",    3.986004418e14 m^3 s^-
2]
sgrA =       BlackHole.new["SgrA*",    1.018e15 m^3 s^-2]
jupiter =    BlackHole.new["Jupiter",  1.268e17 m^3 s^-2]
sun =        BlackHole.new["Sun",      1.32712440018e20 m^3
s^-2]
milkyWay = BlackHole.new["Milky Way", 1.047e22 m^3 s^-2]

// Identify black holes to be displayed
testUnits = [neutrino, electron, protonM, protonR, neutron,
light, bohr, earth, sgrA, jupiter, sun, milkyWay]

for i = 0 to (length[testUnits] - 1)
{
   testUnit = testUnits.popFirst[]

   println["\nParameters of " + testUnit.name[] + " black
hole are... (" + testUnit.howSet[] + ")"]
   println["Luminiferous radius is " + testUnit.radius[]]
```

```
    println["Gravitational parameter is " +
testUnit.gravitationalParameter[]]
    println["Surface tension is " +
testUnit.surfaceTension[]]
    println["Mass is " + testUnit.mass[]]
    if testUnit.name[] != "Light"
    {
        println["Energy is " + testUnit.energy[]]
        println["Frequency is " + testUnit.frequency[]]
    }
    println["Acceleration of gravity at luminiferous radius
is " + testUnit.gravity[testUnit.radius[]]]
    if testUnit.name[] == "Earth"
    {
        println["Acceleration of gravity at surface is " +
testUnit.gravity[6.3781e6 m]]
    }
}
println["\nBohr wavelength = " + Aether.density *
light.gravitationalParameter[]^3 / (4 * pi^2 *
Aether.pressure * protonR.gravitationalParameter[]^2)]
println["What are the implications of the following?"]
println["\nFine structure constant (α) = √(surface tension):
proton to bohr = " + (protonR.surfaceTension[] /
bohr.surfaceTension[])^0.5]
println["Fine structure constant (α) = √(luminiferous
radius): proton to bohr = " + (protonR.radius[] /
bohr.radius[])^0.5]
println["Fine structure constant (α) = √(gravitational
parameter): proton to bohr = " +
(protonR.gravitationalParameter[] /
bohr.gravitationalParameter[])^0.5]
println["Fine structure constant (α) = √(frequency): bohr to
proton = " + (bohr.frequency[] / protonR.frequency[])^0.5]
```

OUTPUT

Parameters of Neutrino black hole are... (radius set using mass)
Luminiferous radius is 1.4199415004421361e-18 m (length)
Gravitational parameter is 0.12761798 m^3 s^-2 (gravitational_parameter)
Surface tension is 2.2773284e+15 s^-2 kg (surface_tension)
Mass is 2.1400000e-37 kg (mass)
Energy is 1.9233361e-20 m^2 s^-2 kg (energy)

Frequency is 3.3602406e+25 s^-1 (frequency)
Acceleration of gravity at luminiferous radius is 6.3295227e+34 m s^-2 (acceleration)

Parameters of Electron black hole are... (radius set using mass)
Luminiferous radius is 2.30124087404205e-16 m (length)
Gravitational parameter is 20.682522 m^3 s^-2 (gravitational_parameter)
Surface tension is 3.6907727e+17 s^-2 kg (surface_tension)
Mass is 9.1093770e-31 kg (mass)
Energy is 8.1870999e-14 m^2 s^-2 kg (energy)
Frequency is 2.0733793e+23 s^-1 (frequency)
Acceleration of gravity at luminiferous radius is 3.9055241e+32 m s^-2 (acceleration)

Parameters of ProtonM black hole are... (radius set using mass)
Luminiferous radius is 2.8179407703040357e-15 m (length)
Gravitational parameter is 253.26389 m^3 s^-2 (gravitational_parameter)
Surface tension is 4.5194656e+18 s^-2 kg (surface_tension)
Mass is 1.6726228e-27 kg (mass)
Energy is 1.5032784e-10 m^2 s^-2 kg (energy)
Frequency is 1.6932028e+22 s^-1 (frequency)
Acceleration of gravity at luminiferous radius is 3.1894041e+31 m s^-2 (acceleration)

Parameters of ProtonR black hole are... (radius directly specified)
Luminiferous radius is 2.8179403227e-15 m (length)
Gravitational parameter is 253.26385 m^3 s^-2 (gravitational_parameter)
Surface tension is 4.5194648e+18 s^-2 kg (surface_tension)
Mass is 1.6726219e-27 kg (mass)
Energy is 1.5032776e-10 m^2 s^-2 kg (energy)
Frequency is 1.6932031e+22 s^-1 (frequency)
Acceleration of gravity at luminiferous radius is 3.1894046e+31 m s^-2 (acceleration)

Parameters of Neutron black hole are... (radius set using mass)
Luminiferous radius is 2.8192373212422198e-15 m (length)
Gravitational parameter is 253.38042 m^3 s^-2 (gravitational_parameter)
Surface tension is 4.5215449e+18 s^-2 kg (surface_tension)
Mass is 1.6749325e-27 kg (mass)
Energy is 1.5053543e-10 m^2 s^-2 kg (energy)
Frequency is 1.6924241e+22 s^-1 (frequency)

Acceleration of gravity at luminiferous radius is 3.1879374e+31 m s^-2 (acceleration)

Parameters of Light black hole are... (radius set using gravitational parameter)
Luminiferous radius is 2.4263102e-12 m (length)
Gravitational parameter is 218065.89 m^3 s^-2 (gravitational_parameter)
Surface tension is 3.8913612e+21 s^-2 kg (surface_tension)
Mass is 1.0676807e-18 kg (mass)
Acceleration of gravity at luminiferous radius is 3.7042057e+28 m s^-2 (acceleration)

Parameters of Bohr black hole are... (radius directly specified)
Luminiferous radius is 5.29177e-11 m (length)
Gravitational parameter is 4.7560058e+6 m^3 s^-2 (gravitational_parameter)
Surface tension is 8.4870386e+22 s^-2 kg (surface_tension)
Mass is 1.1076570e-14 kg (mass)
Energy is 995.51245 m^2 s^-2 kg (energy)
Frequency is 9.0165391e+17 s^-1 (frequency)
Acceleration of gravity at luminiferous radius is 1.6984019e+27 m s^-2 (acceleration)

Parameters of Earth black hole are... (radius set using gravitational parameter)
Luminiferous radius is 0.0044350279 m (length)
Gravitational parameter is 3.9860044e+14 m^3 s^-2 (gravitational_parameter)
Surface tension is 7.1129798e+30 s^-2 kg (surface_tension)
Mass is 6.5206589e+9 kg (mass)
Energy is 5.8604760e+26 m^2 s^-2 kg (energy)
Frequency is 1.0758320e+10 s^-1 (frequency)
Acceleration of gravity at luminiferous radius is 2.0264927e+19 m s^-2 (acceleration)
Acceleration of gravity at surface is 9.7983991 m s^-2 (acceleration)

Parameters of SgrA* black hole are... (radius set using gravitational parameter)
Luminiferous radius is 0.011326777 m (length)
Gravitational parameter is 1.0180000e+15 m^3 s^-2 (gravitational_parameter)
Surface tension is 1.8166094e+31 s^-2 kg (surface_tension)
Mass is 1.0862292e+11 kg (mass)
Energy is 9.7625412e+27 m^2 s^-2 kg (energy)
Frequency is 4.2124474e+9 s^-1 (frequency)
Acceleration of gravity at luminiferous radius is 7.9347831e+18 m s^-2 (acceleration)

Parameters of Jupiter black hole are... (radius set using gravitational parameter)
Luminiferous radius is 1.4108402 m (length)
Gravitational parameter is 1.2680000e+17 m^3 s^-2 (gravitational_parameter)
Surface tension is 2.2627316e+33 s^-2 kg (surface_tension)
Mass is 2.0991134e+17 kg (mass)
Energy is 1.8865890e+34 m^2 s^-2 kg (energy)
Frequency is 3.3819174e+7 s^-1 (frequency)
Acceleration of gravity at luminiferous radius is 6.3703544e+16 m s^-2
(acceleration)

Parameters of Sun black hole are... (radius set using gravitational parameter)
Luminiferous radius is 1476.625 m (length)
Gravitational parameter is 1.3271244e+20 m^3 s^-2 (gravitational_parameter)
Surface tension is 2.3682385e+36 s^-2 kg (surface_tension)
Mass is 2.4066507e+26 kg (mass)
Energy is 2.1629898e+43 m^2 s^-2 kg (energy)
Frequency is 32312.504 s^-1 (frequency)
Acceleration of gravity at luminiferous radius is 6.0865501e+13 m s^-2
(acceleration)

Parameters of Milky Way black hole are... (radius set using gravitational
parameter)
Luminiferous radius is 116494.46 m (length)
Gravitational parameter is 1.0470000e+22 m^3 s^-2 (gravitational_parameter)
Surface tension is 1.8683597e+38 s^-2 kg (surface_tension)
Mass is 1.1817299e+32 kg (mass)
Energy is 1.0620859e+49 m^2 s^-2 kg (energy)
Frequency is 409.577 s^-1 (frequency)
Acceleration of gravity at luminiferous radius is 7.7150038e+11 m s^-2
(acceleration)

Bohr wavelength = 4.556335e-8 m (length)

What are the implications of the following?
Fine structure constant (α) = $\sqrt{}$(surface tension): proton to bohr =
0.007297354041020622
Fine structure constant (α) = $\sqrt{}$(luminiferous radius): proton to bohr =
0.007297353972502636
Fine structure constant (α) = $\sqrt{}$(gravitational parameter): proton to bohr =
0.007297353972502636
Fine structure constant (α) = $\sqrt{}$(frequency): bohr to proton =
0.007297353903984649

Electromagnetism.frink

```
// Defines electromagnetic units

use Aether.frink
use BlackHole.frink

class Electromagnetism
{
   // Create BlackHole objects
   class var electron = BlackHole.new["Electron",
9.1093837015e-31 kg]
   class var proton  =  BlackHole.new["Proton",
1.67262192369e-27 kg]

   // Reciprocal of gravitational parameter of proton
   class rawElectricConstant[] :=  Aether.density /
proton.surfaceTension[]

   // Luminiferous radius of proton
   class rawMagneticConstant[] :=  proton.surfaceTension[] /
Aether.pressure

   class electricConstant[] := rawElectricConstant[] *
electron.chiFactor[]

   class magneticConstant[] := rawMagneticConstant[] /
electron.chiFactor[]

   class chiFactor[] := electron.chiFactor[]

   new[] :=
   {
   }
}
```

Electromagnetism.frink demonstrates how the charge squared per electron (Eq 5-1) determines the electric (Eq 5–2) and magnetic (Eq 5–3) constants. The following test program demonstrates how to use the Electromagnetism class.

ElectromagnetismTest.frink

```
// Tests the Electromagnetism class
```

```
use Aether.frink
use Electromagnetism.frink

testUnit = Electromagnetism.new[]

println["Chi factor is " + testUnit.chiFactor[]]
println[""]
println["Raw electric constant is 1/" +
1/testUnit.rawElectricConstant[]]
println["Raw magnetic constant is " +
testUnit.rawMagneticConstant[]]
println[""]
println["Electric constant is " +
testUnit.electricConstant[]]
println["Magnetic constant is " +
testUnit.magneticConstant[]]
```

OUTPUT

Chi factor is 2.2424472e-9 s^2 kg^-1 A^2 (charge_squared_per_mass)

Raw electric constant is 1/253.26389 m^3 s^-2 (gravitational_parameter)
Raw magnetic constant is 2.8179408e-15 m (length)

Electric constant is 8.8541922e-12 m^-3 s^4 kg^-1 A^2 (permittivity)
Magnetic constant is 0.0000012566364 m s^-2 kg A^-2 (permeability)

Light.frink

```
// Defines light characteristics

use Aether.frink
use BlackHole.frink

class Light
{
    class var wavelengthBalmer = 3.6450682e-7 m
    class var electron = BlackHole.new["Electron",
9.10938356e-31 kg]
    class var light = BlackHole.new["Light",
2.18065888937262e5 m^3 s^-2]
    var myColor
    var myM
```

```
    var myN
    var myWavelength

    class color[] := myColor

    class m[] := myM

    class n[] := myN

    class energy[] := pressure[] * electron.volume[]

    class pressure[] := light.surfaceTension[] / myWavelength

    class wavelength[] := myWavelength

    // Constructor for name, m, & n
    new[m, n, objColor is string] :=
    {
        myColor = objColor
        myM = m
        myN = n
        myWavelength = wavelengthBalmer * n^2 / (n^2 - m^2)
    }

    // Constructor for name and wavelength
    new[objWavelength is length, objColor is string] :=
    {
        myColor = objColor
        myWavelength = objWavelength
    }
}
```

Light.frink demonstrates how the Balmer equation (Eq 8–1) determines the wavelength of light, and how that determines light's pressure (Eq 6–3). The following test program demonstrates how to use the Light class.

LightTest.frink

```
// Tests the Light class

use Light.frink

setPrecision[6]
```

```
// Create light objects
m8n12  = Light.new[8, 12, "red"]
m8n13  = Light.new[8, 13, "yellow"]
m8n14  = Light.new[8, 14, "lime"]
m8n15  = Light.new[8, 15, "green"]
m8n16  = Light.new[8, 16, "cyan"]
m8n17  = Light.new[8, 17, "teal"]
m8n18  = Light.new[8, 18, "indigo"]
m8n19  = Light.new[8, 19, "blue"]
m8n20  = Light.new[8, 20, "bluer"]
m8n21  = Light.new[8, 21, "blue 2"]
m8n22  = Light.new[8, 22, "blue 3"]
m8n23  = Light.new[8, 23, "blue 4"]
m8n24  = Light.new[8, 24, "violet"]
Balmer = Light.new[0,  1, "Balmer"]
Bohr   = Light.new[Balmer.wavelength[]/8, "Bohr"]

testUnits = [m8n12, m8n13, m8n14, m8n15, m8n16, m8n17,
m8n18, m8n19, m8n20, m8n21, m8n22, m8n23, m8n24]

for i = 0 to (length[testUnits]-1)
{
   testUnit = testUnits.popFirst[]
   println["m<n=(" + testUnit.m[] + "," + testUnit.n[] +
") \u03bb=" + testUnit.wavelength[] + " P=" +
testUnit.pressure[] + " (" + testUnit.color[] + ")"]
   testUnits.push[testUnit]
}
println["m<n=(0, 1) \u03bb=" + Balmer.wavelength[] + " P=" +
Balmer.pressure[] + " (" + Balmer.color[] + ")"]
println["                \u03bb=" + Bohr.wavelength[] + " P=" +
Bohr.pressure[] + " (" + Bohr.color[] + ")"]
println[""]

for i = 0 to (length[testUnits]-1)
{
   testUnit = testUnits.popFirst[]
   println["m<n=(" + testUnit.m[] + "," + testUnit.n[] +
") \u03bb=" + testUnit.wavelength[] + " E=" +
testUnit.energy[] + " (" + testUnit.color[] + ")"]
}
```

```
println["m<n=(0, 1) \u03bb=" + Balmer.wavelength[] + " E=" +
Balmer.energy[] + " (" + Balmer.color[] + ")"]
println["            \u03bb=" + Bohr.wavelength[] + " E=" +
Bohr.energy[] + " (" + Bohr.color[] + ")"]
```

OUTPUT

m<n=(8,12) λ=6.56113e-7 m (length) P=5.93095e+27 m^-1 s^-2 kg (pressure) (red)
m<n=(8,13) λ=5.86681e-7 m (length) P=6.63284e+27 m^-1 s^-2 kg (pressure) (yellow)
m<n=(8,14) λ=5.41238e-7 m (length) P=7.18977e+27 m^-1 s^-2 kg (pressure) (lime)
m<n=(8,15) λ=5.09406e-7 m (length) P=7.63903e+27 m^-1 s^-2 kg (pressure) (green)
m<n=(8,16) λ=4.86008e-7 m (length) P=8.00681e+27 m^-1 s^-2 kg (pressure) (cyan)
m<n=(8,17) λ=4.68187e-7 m (length) P=8.31158e+27 m^-1 s^-2 kg (pressure) (teal)
m<n=(8,18) λ=4.54230e-7 m (length) P=8.56697e+27 m^-1 s^-2 kg (pressure) (indigo)
m<n=(8,19) λ=4.43055e-7 m (length) P=8.78306e+27 m^-1 s^-2 kg (pressure) (blue)
m<n=(8,20) λ=4.33938e-7 m (length) P=8.96758e+27 m^-1 s^-2 kg (pressure) (bluer)
m<n=(8,21) λ=4.26386e-7 m (length) P=9.12639e+27 m^-1 s^-2 kg (pressure) (blue 2)
m<n=(8,22) λ=4.20051e-7 m (length) P=9.26403e+27 m^-1 s^-2 kg (pressure) (blue 3)
m<n=(8,23) λ=4.14674e-7 m (length) P=9.38416e+27 m^-1 s^-2 kg (pressure) (blue 4)
m<n=(8,24) λ=4.10070e-7 m (length) P=9.48953e+27 m^-1 s^-2 kg (pressure) (violet)
m<n=(0, 1) λ=3.64507e-7 m (length) P=1.06757e+28 m^-1 s^-2 kg (pressure) (Balmer)
             λ=4.55634e-8 m (length) P=8.54055e+28 m^-1 s^-2 kg (pressure) (Bohr)

m<n=(8,12) λ=6.56113e-7 m (length) E=3.02680e-19 m^2 s^-2 kg (energy) (red)
m<n=(8,13) λ=5.86681e-7 m (length) E=3.38500e-19 m^2 s^-2 kg (energy) (yellow)
m<n=(8,14) λ=5.41238e-7 m (length) E=3.66922e-19 m^2 s^-2 kg (energy) (lime)
m<n=(8,15) λ=5.09406e-7 m (length) E=3.89849e-19 m^2 s^-2 kg (energy) (green)
m<n=(8,16) λ=4.86008e-7 m (length) E=4.08619e-19 m^2 s^-2 kg (energy) (cyan)
m<n=(8,17) λ=4.68187e-7 m (length) E=4.24172e-19 m^2 s^-2 kg (energy) (teal)
m<n=(8,18) λ=4.54230e-7 m (length) E=4.37206e-19 m^2 s^-2 kg (energy) (indigo)
m<n=(8,19) λ=4.43055e-7 m (length) E=4.48234e-19 m^2 s^-2 kg (energy) (blue)

m<n=(8,20) λ=4.33938e-7 m (length) E=4.57651e-19 m^2 s^-2 kg (energy) (bluer)

m<n=(8,21) λ=4.26386e-7 m (length) E=4.65755e-19 m^2 s^-2 kg (energy) (blue 2)

m<n=(8,22) λ=4.20051e-7 m (length) E=4.72780e-19 m^2 s^-2 kg (energy) (blue 3)

m<n=(8,23) λ=4.14674e-7 m (length) E=4.78910e-19 m^2 s^-2 kg (energy) (blue 4)

m<n=(8,24) λ=4.10070e-7 m (length) E=4.84288e-19 m^2 s^-2 kg (energy) (violet)

m<n=(0, 1) λ=3.64507e-7 m (length) E=5.44823e-19 m^2 s^-2 kg (energy) (Balmer)

λ=4.55634e-8 m (length) E=4.35858e-18 m^2 s^-2 kg (energy) (Bohr)

B Fractals

This appendix summarizes and slightly extends Chapter 4 of the author's first book and also revises it to use the Frink programming language.[58] Fractal geometry transforms chaos into order, and may well transform random fluctuations of the aether into reality as we know it.

The book "The Fractal Geometry of Nature" by Benoit Mandelbrot, meticulously explains how predetermined outcomes, including the geometry of Nature, predictably emerges from totally random occurrences. Half a century ago, Mandelbrot thought his computer malfunctioned when it generated order from chaos. Be advised, his book is for math lovers only.

Interested readers will find a wealth of information by researching the Mandelbrot Set. The cover of this book is a typical rendering of a small portion of the Mandelbrot Set.[59] Could that be the recurring pattern of neutrons, protons, electrons, filaments between galaxies, and everything else in the universe?

Like Appendix A, to avoid confusion over line wrapping, this appendix emulates retro green and white printer paper. To reproduce the cover fractal, go to the https://math.hws.edu/eck/js/mandelbrot/MB.html web page, enter the following parameters, and then press the "Show XML Import/Export" button:

```
<?xml version='1.0'?>
<mandelbrot_settings_2>
<image_size width='6000' height='4500'/>
```

---

58 Heffron, Michael. *Ramblings of an Old Man: about prime number seventeen* (self-published via Kindle Direct Publishing, 2020).

59 The black blobs represent points that belong to the Mandelbrot Set (the set of complex numbers c for which the function $f_c(z) = z^2 + c$ does not diverge to infinity when iterated from z=0). The colors represent how quickly points diverge to infinity when they are not part of the Mandelbrot Set.

```
<limits>
    <xmin>-0.0.472148889249369302905</xmin>
    <xmax>-0.0.460501789121007407624</xmax>
    <ymin>0.587974922948997294658</ymin>
    <ymax>0.596702565261542884018</ymax>
</limits>
<palette colorType='HSB'>
    <divisionPoint position='0' color='0.15730337078651685;
0.6409574468085106;1'/>
    <divisionPoint position='1' color='1.1573033707865168;
0.6409574468085106;1'/>
</palette>
<palette_mapping length='241' offset='0'/>
<max_iterations value='10000'/>
</mandelbrot_settings_2>
```

For readers who want less math, only want to know a little more, and/or don't want to read a book about it–this appendix provides additional details. It contains a few examples of fractals generated using the Iterated Function System described in the book "Fractals Everywhere" by Michael Barnsley. If your interest is far more complex fractals, which can reproduce photograph quality images randomly, you might want to read the book "Fractal Image Compression" by Michael Barnsley and Lyman Hurd.

The IteratedFunctionSystem.frink module in this appendix generated all of the fractal images, which immediately follow each corresponding array of coefficients. The InteratedFunctionSystem.frink module creates a graph window. Next it iterates for the specified number of kilosteps, over however many rows of coefficients were passed into the drawIFS function, randomly selecting which coefficient to apply to the last point on the graph to calculate where the next point should be on the graph. That doesn't sound like it should work, does it?

In his book "Fractals Everywhere," Michael Barnsley describes how the following "affine transform" relocates, resizes, rotates, and tilts geometric shapes.

Eq B-1 $F(x,y) = \begin{bmatrix} a & b \\ c & d \end{bmatrix} \begin{bmatrix} x \\ y \end{bmatrix} + \begin{bmatrix} e \\ f \end{bmatrix} = (ax + by + e, cx + dy + f)$

The iterated function system assigns probabilities to a set of such functions, then randomly selects one function from the set at each iteration.

Examine the comments for the CHAOS fractal (a few pages ahead) to understand how a fractal combines many such functions to produce order from chaos.

IteratedFunctionSystem.frink

```
// The following Frink program implements an Iterated
Function System
// fractal image generator, rendered in a different color
for each
// row of the coefficients.

drawIFS[title is string, coeff is array, k_steps = 30,
doFile = false] :=
{
    // Find the probability for each row of the coefficients
array
    prob = new array
    sum = 0.0
    for n = rangeOf[coeff]
    {
        sum = sum + coeff@n@6
        prob.push[sum]
    }

    // Initialize algorithm parameters
    iter = 0
    steps = 1000 * k_steps
    if k_steps < 10
       steps = 10000
    if k_steps > 250
       steps = 250000
    x = 1.0
    y = 1.0

    // Create graph window with light gray background.
    g = new graphics
    g.backgroundColor[0.9,0.9,0.9]
    g.color[0, 0, 0]

    // Use a different color for each row of the affine
transform.
    colors = [g.color[0,0,0],   // Black
              g.color[1,0,0],   // Red
              g.color[0,1,0],   // Green
              g.color[0,0,1],   // Blue
              g.color[1,0,1],   // Magenta
```

```
            g.color[0,1,1],   // Cyan
            g.color[1,.7,0]] // Yellowish
    win = g.show[]

    // Iterate for the specified number of steps.
    OUTERLOOP:
    while iter < steps
    {
        iter = iter + 1
        // Update display every 500 iterations.
        if (iter % 500)==0
            win.repaint[]
        // Randomly select which row of coefficients to use.
        rand = randomFloat[0.0, 1.0]
        for n = rangeOf[coeff]
        {
            if rand <= prob@n
            {
                x0 = x
                y0 = y
                x = coeff@n@0*x0 + coeff@n@1*y0 + coeff@n@4
                y = coeff@n@2*x0 + coeff@n@3*y0 + coeff@n@5
                // Allow 1000 iterations for attractor to take
effect.
                if iter >= 1000
                {
                    i = n % length[colors]
                    g.color[colors@i]
                    g.drawRectSize[x,-y,.001,.001]
                    next OUTERLOOP
                }
            }
        }
    }
    // After last iteration, change window background to
white.
    g.backgroundColor[1,1,1]
    win.repaint[]
    if doFile
    {
        g.show[]
        g.write[title+".jpg",1500,undef]
```

```
    }
    return g
}
```

IteratedFunctionSystemTest.frink

```
// Test of IteratedFunctionSystem class.

use IteratedFunctionSystem.frink

// Bourke Tree
//            a        b        c        d        e        f      p
Bourke = [[ 0.1950,-0.4880, 0.3440, 0.4430, 0.4431, 0.2452, 0.2],
          [ 0.4620, 0.4140,-0.2520, 0.3610, 0.2511, 0.5692, 0.2],
          [-0.6370, 0.0000, 0.0000, 0.5010, 0.8562, 0.2512, 0.2],
          [-0.0350, 0.0700,-0.4690, 0.0220, 0.4884, 0.5069, 0.2],
          [-0.0580,-0.0470, 0.4530,-0.1110, 0.5976, 0.0969, 0.2]]
```

```
// Chaos: (<) left, (^) top, (>) right, (v) bottom, (-) center ... of letter
//            a            b            c            d       e      f      p
Chaos=[[ 0.0000000, 0.0530000,-0.4290000, 0.0000000,-7.083, 5.430, 0.0526],//<C
        [ 0.1430000, 0.0000000, 0.0000000,-0.0530000,-5.619, 8.513, 0.0526],//^C
        [ 0.1430000, 0.0000000, 0.0000000, 0.0830000,-5.619, 2.057, 0.0527],//vC
        [ 0.0000000, 0.0530000, 0.4290000, 0.0000000,-3.952, 5.430, 0.0526],//<H
        [ 0.1190000, 0.0000000, 0.0000000, 0.0530000,-2.555, 4.536, 0.0526],//-H
        [-0.0123806,-0.0649723, 0.4238190, 0.0018980,-1.226, 5.235, 0.0527],//>H
        [ 0.0852291, 0.0506328, 0.4204490, 0.0156626,-0.421, 4.569, 0.0526],//<A
        [ 0.1044320, 0.0052912, 0.0570516, 0.0527352, 0.976, 8.113, 0.0526],//^A
        [-0.0081419,-0.0417935, 0.4239220, 0.0041597, 1.934, 5.370, 0.0527],//>A
        [ 0.0930000, 0.0000000, 0.0000000, 0.0530000, 0.861, 4.536, 0.0526],//-A
        [ 0.0000000, 0.0530000,-0.4290000, 0.0000000, 2.447, 5.430, 0.0526],//<O
        [ 0.1190000, 0.0000000, 0.0000000,-0.0530000, 3.363, 8.513, 0.0527],//^O
        [ 0.1190000, 0.0000000, 0.0000000, 0.0530000, 3.363, 1.487, 0.0526],//vO
        [ 0.0000000, 0.0530000, 0.4290000, 0.0000000, 3.972, 4.569, 0.0526],//>O
        [ 0.1239980,-0.0018396, 0.0006912, 0.0629731, 6.275, 7.716, 0.0527],//^S
        [ 0.0000000, 0.0530000, 0.1670000, 0.0000000, 5.215, 6.483, 0.0526],//<S
        [ 0.0710000, 0.0000000, 0.0000000, 0.0530000, 6.279, 5.298, 0.0526],//-S
        [ 0.0000000,-0.0530000,-0.2380000, 0.0000000, 6.805, 3.714, 0.0527],//>S
        [-0.1210000, 0.0000000, 0.0000000, 0.0530000, 5.941, 1.487, 0.0526]]//vS
```

Frink software generated this image of the legible words CHAOS completely randomly. Each time you run the module, it randomly plots the points in a different order–but eventually always generates a legible image of the word CHAOS made up of smaller words CHAOS (which are made of even smaller words CHAOS, down to the minimum resolution of the monitor). A good programmer can modify the software to zoom in even beyond the monitor's resolution by adding positioning and a magnification factor. That is left as an exercise for interested readers!

Those readers who take the time to understand how and why fractals work may begin to understand how the fractal universe emerges from the aether, guided by the laws of physics.

```
// Fern
//         a       b       c      d      e      f      p
Fern = [[ 0.00,  0.00,  0.00, 0.16, 0.0, 0.00, 0.01],
        [ 0.85,  0.04, -0.04, 0.85, 0.0, 1.60, 0.85],
        [ 0.20, -0.26,  0.23, 0.22, 0.0, 1.60, 0.07],
        [-0.15,  0.28,  0.26, 0.24, 0.0, 0.44, 0.07]]
```

```
// Galaxy
//            a      b      c      d      e      f      p
Galaxy = [[-0.08,  0.06,  0.00, -0.28, -2.02, -2.40, 0.05],
          [-0.05,  0.20, -0.03, -0.21, -1.51, -3.46, 0.05],
          [-0.05,  0.23, -0.05, -0.20,  2.04,  2.47, 0.05],
          [-0.01,  0.30, -0.05, -0.04,  0.67,  3.18, 0.05],
          [ 0.70,  0.52, -0.53,  0.69,  0.54,  0.14, 0.80]]
```

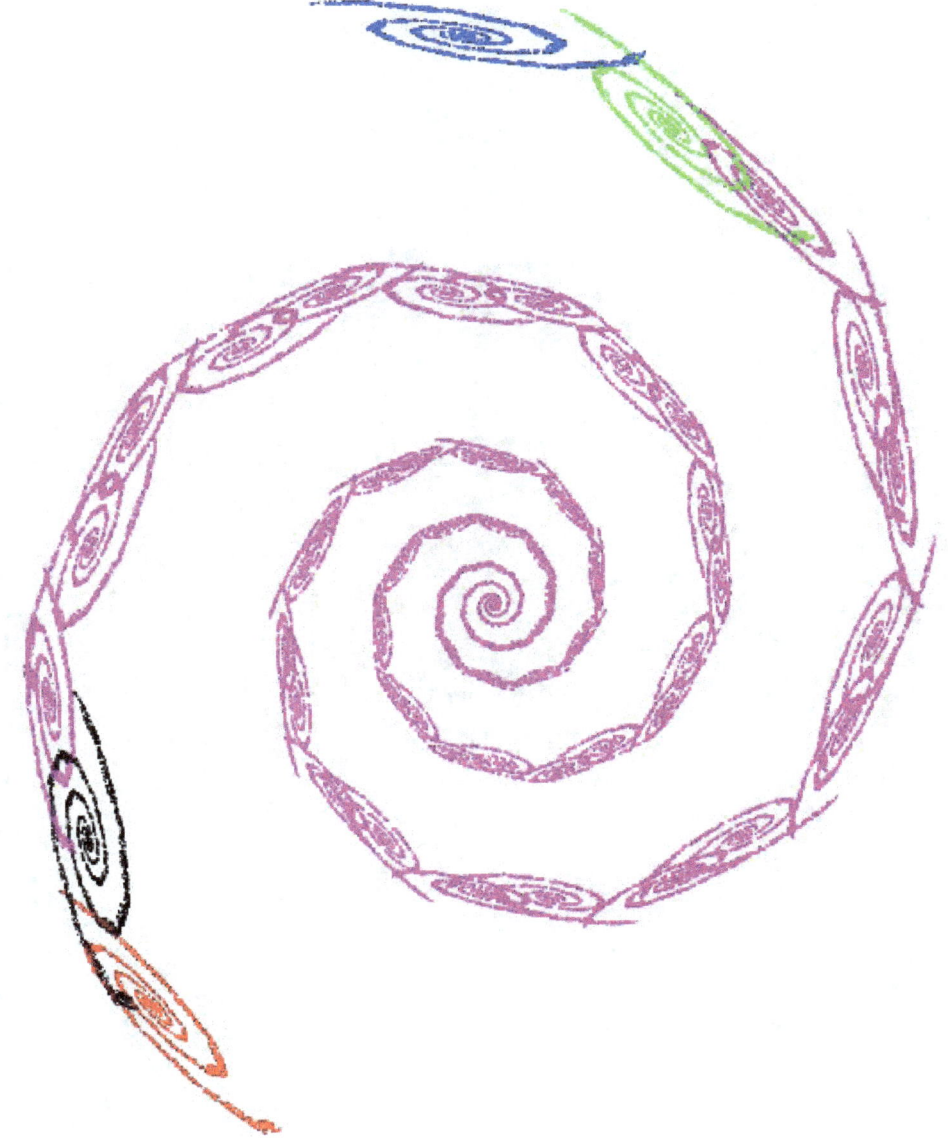

```
// Maple Leaf
//               a       b       c      d       e       f      p
MapleLeaf = [[0.14,   0.01,   0.00, 0.51,  -0.08,  -1.31, 0.10],
            [0.43,   0.52,  -0.45, 0.50,   1.49,  -0.75, 0.35],
            [0.45,  -0.49,   0.47, 0.47,  -1.62,  -0.74, 0.35],
            [0.49,   0.00,   0.00, 0.51,   0.02,   1.62, 0.20]]
```

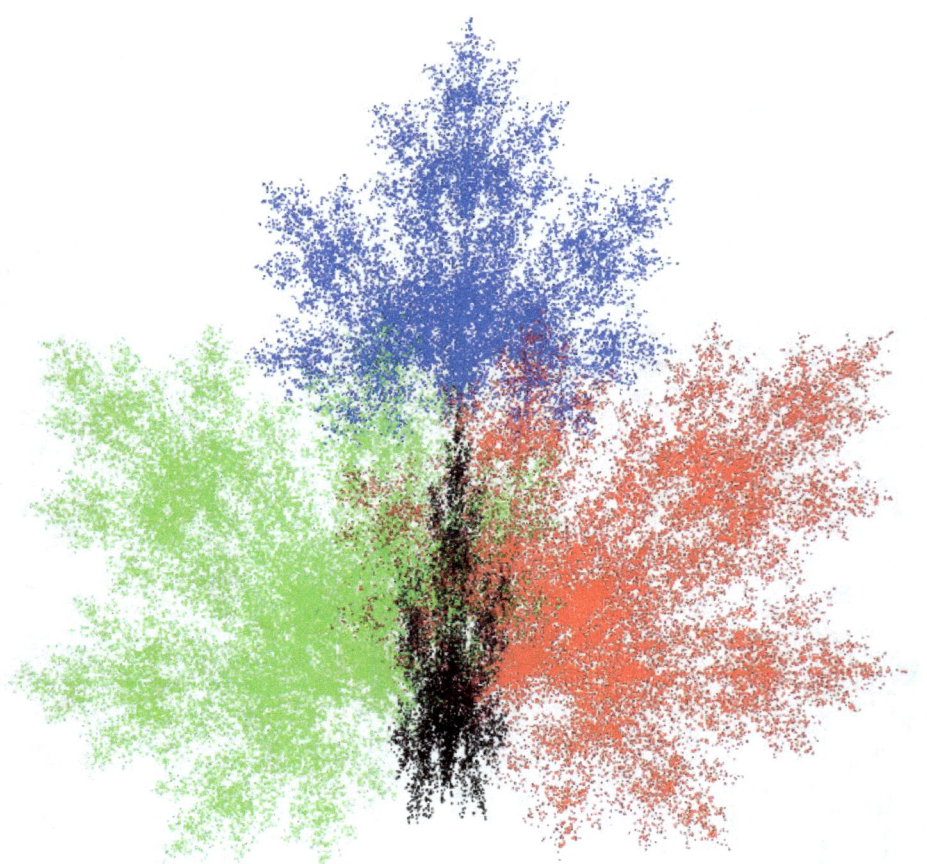

```
// Sierpinski Triangle (p-adic fans take note!)
//           a    b    c    d     e     f     p
Triangle = [[0.5, 0.0, 0.0, 0.5,  1.0,  1.0, 0.33],
            [0.5, 0.0, 0.0, 0.5,  1.0, 50.0, 0.33],
            [0.5, 0.0, 0.0, 0.5, 50.0, 50.0, 0.34]]
```

```
// Spiral
//              a        b        c        d        e        f      p
Spiral = [[0.7517,-0.2736, 0.2736, 0.7517, 0.0000, 0.000, 0.7],
          [0.2000, 0.0000, 0.0000, 0.2000, 1.0000,-0.364, 0.1],
          [0.2000, 0.0000, 0.0000, 0.2000,-0.3640, 1.000, 0.1],
          [0.2000, 0.0000, 0.0000, 0.2000,-0.7280,-0.728, 0.1]]
```

```
// Vortex
//           a       b       c       d       e       f       p
Vortex = [[ 0.4426,-0.8411, 0.7244, 0.5387, 14.9191, 9.3286, 0.9127],
          [-0.4242,-0.0652,-0.1758,-0.2182,  3.8096, 6.7415, 0.0873]]
```

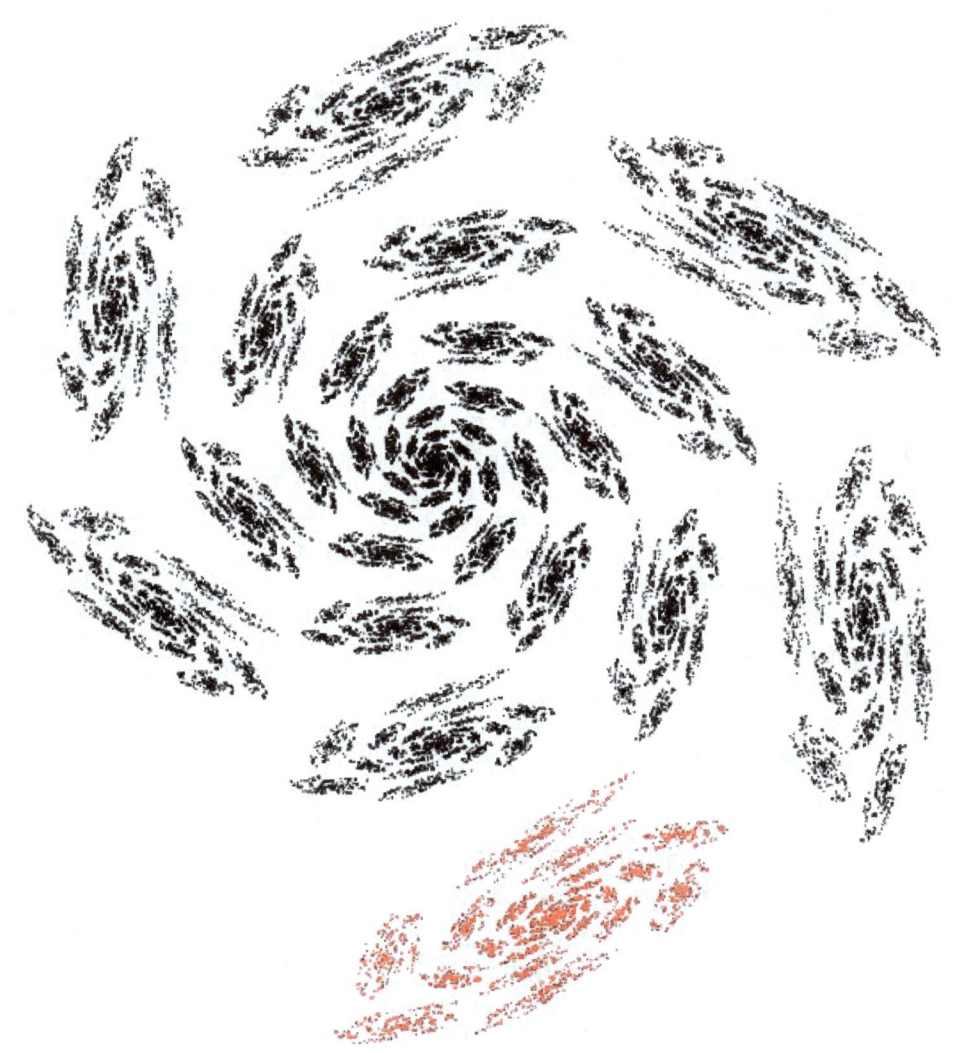

```
whichTest = ""

do
{
    whichTest = input["Run what test ?\n b = Bourke tree, c =
chaos, f = fern, o = order,\n m = maple leaf, s = spiral, t
= triangle, v = vortex, OK = exit (add F to generate a
file)"]

    if uppercase[whichTest]=="B"
        g = drawIFS["BourkeTree", Bourke, 60]
    if uppercase[whichTest]=="BF"
        g = drawIFS["BourkeTree", Bourke, 60, true]
    if uppercase[whichTest]=="C"
        g = drawIFS["Chaos", Chaos, 60]
    if uppercase[whichTest]=="CF"
        g = drawIFS["Chaos", Chaos, 60, true]
    if uppercase[whichTest]=="F"
        g = drawIFS["Fern", Fern, 40]
    if uppercase[whichTest]=="FF"
        g = drawIFS["Fern", Fern, 120, true]
        if uppercase[whichTest]=="G"
        g = drawIFS["Galaxy", Galaxy, 40]
    if uppercase[whichTest]=="GF"
        g = drawIFS["Galaxy", Galaxy, 120, true]
    if uppercase[whichTest]=="M"
        g = drawIFS["Maple Leaf", MapleLeaf, 100]
    if uppercase[whichTest]=="MF"
        g = drawIFS["Maple Leaf", MapleLeaf, 100, true]
    if uppercase[whichTest]=="O"
        g = drawIFS["Order", Order, 30]
    if uppercase[whichTest]=="OF"
        g = drawIFS["Order", Order, 30, true]
    if uppercase[whichTest]=="S"
        g = drawIFS["Spiral", Spiral, 30]
    if uppercase[whichTest]=="SF"
        g = drawIFS["Spiral", Spiral, 30, true]
    if uppercase[whichTest]=="T"
        g = drawIFS["SierpinskiTriangle", Triangle, 50]
    if uppercase[whichTest]=="TF"
        g = drawIFS["SierpinskiTriangle", Triangle, 200, true]
    if uppercase[whichTest]=="V"
```

```
        g = drawIFS["Vortex", Vortex, 50]
    if uppercase[whichTest]=="VF"
        g = drawIFS["Vortex", Vortex, 100, true]
    if whichTest==""
        whichTest = "X"
} while (uppercase[whichTest] != "X")
```

Are any brave matrix math experts willing to try three dimensional affine transforms?

$$F(x,y,z) = \begin{bmatrix} a & b & c \\ d & e & f \\ g & h & i \end{bmatrix} \begin{bmatrix} x \\ y \\ z \end{bmatrix} + \begin{bmatrix} j \\ k \\ l \end{bmatrix} = \cdots$$

It gets very complicated... in a way that only matrix mathematicians can learn to love!

Now ponder this!

Imagine a vast sea of tranquil aether in perfect equilibrium. If not all "particles" of aether are the same size, could those odd particles be the coefficients of the fractal universe, and could their rate of encounter be the probability? What impact do clumps of matter within the aether have upon the surrounding aether flow?

C DERIVATION OF EQUATIONS

WARNING: this entire appendix is a math box, so it omits the box and gray tint. This appendix provides background material and supplemental information for many equations.

Chapter 3 of this book emerged from the simple answer to the question "If aether does exist, what does it look like?" Much of the remainder of this book refactors equations from the author's prior book to demonstrate how the aether explains mass, energy, gravity, electricity, magnetism, and light.[60]

Earlier drafts of this book explained everything from whatever frame of reference the prior book used, just refactored as necessary to account for the role played by density and pressure of the aether. Background material for some of the following equations includes those original equations.

After finding the connection between surface tension and the gravitational parameter, the author eventually realized that everything is much easier to explain in terms of surface tension. That realization led to refactoring the entire book to explain how surface tension governs the universe.

That being said, **momentum (kg·m/s) is the most important concept in all of physics!** Neither surface tension (kg/s^2) nor anything else would exist without momentum.

1.1 momentum

Eq 1-1
$$m_1 u_1 + \cdots + m_n u_n = m_1 v_1 + \cdots + m_n v_n$$

expresses the conservation of momentum, where $m_\#$ represents the mass of object #, $u_\#$ represents the pre-collision velocity of object #, $v_\#$ represents the

---

60 Heffron, Michael. *Ramblings of an Old Man: about the scientific method* (self-published via Kindle Direct Publishing, 2020).

post-collision velocity of object #, n represents the number of objects involved in the collision, and … denotes unlisted terms for objects 2 through n-1.

Note that objects may fragment or merge during inelastic collisions. When a collision is elastic, all colliding objects exit without fragmenting or merging.

2.0   density and pressure

Pressure $(kg/m \cdot s^2)$ is energy $(kg \cdot m^2/s^2)$ per volume (m^3), where the energy is alternatively momentum $(kg \cdot m/s)$ times velocity (m/s).

Density (kg/m^3) is mass (kg) per volume (m^3), where the mass (kg) is momentum $(kg \cdot m/s)$ per velocity (m/s).

Comparing the units of surface tension (kg/s^2) with those of pressure $(kg/m \cdot s^2)$, reveals that pressure (P) is surface tension per distance, $P = \frac{\gamma}{d}$, where γ is surface tension, and d is distance.

2.1   speed of sound

The propagation speed of the energy redistribution by sound waves is

Eq 2-1
$$c = \sqrt{\frac{P}{\rho}},$$

where P is the pressure of the medium (such as air) through which the sound waves propagate, and ρ is the density of the medium. The letter c represents speed (from the Latin word *celeritas*, meaning velocity).

2.2   Boyle's Law

Boyle's Law essentially states that

Eq 2-2
$$\frac{P_1 V_1}{T_1} = \frac{P_2 V_2}{T_2},$$

where P_1 is the initial pressure V_1 is the initial volume, T_1 is the initial temperature, P_2 is the final pressure, V_2 is the final volume, and T_2 is the final temperature. That means when one of those variables changes, one or both of the others must also change to prevent the ratio (of potential energy to kinetic energy) from changing.

It is common to view force times distance as kinetic energy, and pressure (force per area) times volume as potential energy.

The units of pressure are $kg/m \cdot s^2$. Multiplication by volume (m^3), transforms the view from pressure into potential energy $(kg \cdot m^2/s^2)$. Kinetic energy (work) has the same units $(kg \cdot m^2/s^2)$, but is often viewed as force $(kg \cdot m/s^2)$ times distance (m). Continue to remain focused on the fact that those are different views of the same molecules.

3.1   speed of light

Conveniently, c has already come to represent the speed of light. If we let P represent the pressure of the aether, and ρ represent the density of the aether–then the equation for the speed of sound (Eq 2-1),

Eq 3-1 $$c = \sqrt{\frac{P}{\rho}},$$

works equally well to find the speed of light waves in the aether.

3.2   pressure is density times the speed of light

Rearranging Eq 3–1 to solve for pressure,

Eq 3-2 $$P = \rho c^2$$

is a general equation that applies everywhere throughout the vast expanse of the aether. Thus, it is reasonable to view the pressure of the aether as whatever constitutes its density colliding with itself at the speed of light.

Aether.frink in Appendix A computes the density, mass flux, and pressure of the aether. Note that for any velocity (v), $P = \rho v^2$ (not just for the speed of light).

Density (ρ, kg/m³) times velocity (v, m/s) is mass flux (φ, kg/m²s), which is momentum per volume. Mass flux times velocity is pressure (P, kg/m·s²), which means pressure per velocity is mass flux, which colliding with itself is equal to density times pressure, $\varphi^2 = \rho P$, $P = \varphi c$, $\rho = \dfrac{\varphi}{c}$. Note that mass flux times volume is momentum.

3.3   shape factor

Let § represent a shape factor for subatomic particles composed of the aether. Those shapes are best visualized as amorphous blobs; however, since surface tension often causes the particles to behave like spheres, substituting the scalar multiplier $\frac{4}{3}\pi$ (to find the volume of a sphere) for § is often a very good approximation. For example, $\S R^3 \approx \frac{4}{3}\pi R^3$, where R is the radius of the sphere. The volume of an ellipsoid $\left(\frac{4}{3}\pi ABC\right)$ is also a good approximation–where A, B, and C are the lengths of the semi-axes of the ellipsoid (spherical when $A=B=C$).

3.3   mass is density times volume

For any given particle of radius R, density (ρ) of the aether determines the mass,

Eq 3-3 $$m = \rho \S R^3.$$

3.4 energy is pressure times volume

For any given particle of radius R, pressure (P) of the aether determines the energy,

Eq 3-4
$$E = P\S R^3.$$

$E = mc^2$ is equivalent to $E = m\left(\dfrac{P}{\rho}\right)$. Since mass is density times volume ($m = \rho\S R^3$), then $E = m\left(\dfrac{P}{\rho}\right) = \rho\S R_E^3\left(\dfrac{P}{\rho}\right)$. After canceling the densities, the energy is $E = \S R^3 P = P\S R^3$. That is, the $E = mc^2$ energy equals the pressure of the aether **(without regard to density)** times the volume of the black hole.

Thus, for any given particle of radius R, density of the aether (ρ) determines its mass ($\rho\S R^3$), and pressure of the aether (P) determines its "relativistic" energy ($P\S R^3$). BlackHole.frink in Appendix A computes mass and energy by those methods.

3.5 relativistic energy

Pressure times volume is energy. Density times volume is mass. Thus, multiplying each side of the equation $P = \rho c^2$ by a specific volume produces Einstein's much more familiar energy equation,

Eq 3-5
$$E = mc^2,$$

where E is the energy released, m is the mass of a particle, and c is the speed of light. That is the same energy as Eq 3-4, for the luminiferous radius (R) of a particle of mass m. Notice that although $P = \rho c^2$ is a general equation that applies throughout the aether (including to that associated with mass and energy), the equation $E = mc^2$ only applies to the energy and mass of the aether confined within the volume of the mass.

Rearranging Eq 3-3 to find the radius from the mass yields

3.6 radius determined from mass

Any given particle of mass (m) has a luminiferous radius (R) of

Eq 3-6
$$R = \sqrt[3]{\frac{m}{\rho\S}},$$

where ρ is the density of the aether. The aethereal view is that R is the luminiferous radius for the whirling sphere that contains the specified mass composed of the density of the aether.

4.1   gravitational parameter

Astronomer's typically use the Greek letter Mu (μ) to represent the "standard" gravitational parameter; however, due to problems with the astronomical definition, and to avoid confusion with the magnetic constant (μ_0) presented in the next chapter, this book substitutes the ancient Greek letter Koppa ($\varkappa$) to define the gravitational parameter based upon Kepler's Third Law. Thus,

Eq 4-1 $\varkappa = v^2 r,$

where v is the velocity of the satellite, and r is the radius of its orbit.

4.1a   Schwarzschild radius

Chapter 6 of the author's prior book[61] explains that Newton's gravitational constant (G) is not quite correct, and so it is inaccurate to define the Schwarzschild radius as

Eq 4-1a $R_S = \dfrac{2GM}{c^2},$

where M is the mass of the black hole, and c is the speed of light.

4.1b   luminiferous radius

Consequently, this book substitutes the more accurate definition of the luminiferous radius (R_L) of a black hole, which is

Eq 4-1b $R_L = \dfrac{v^2 r}{c^2},$

where c is the speed of light, v is the velocity of a satellite in orbit, and r is the radius of that orbit. That is to say $c^2 R_L = v^2 r.$[62]

4.2   gravitational parameter due to surface tension

The gravitational parameter ($\varkappa = v^2 r$) of any given black hole of luminiferous radius R_L is

Eq 4-2 $\varkappa = v^2 r = c^2 R_L = \dfrac{P R_L}{\rho},$

---

61 Heffron, Michael. *Ramblings of an Old Man: about the scientific method*. (self-published via Kindle Direct Publishing, 2020). Chapter 6 (page 43) explains that the formal definition of the Schwarzschild radius isn't quite right, because it relies upon Newton's not quite correct Law of Gravitation (the error of which Chapter 5 explains in great detail).

62 Heffron, Michael. *Ramblings of an Old Man: about the scientific method* (self-published via Kindle Direct Publishing, 2020). Chapter 7 (page 52).

where v is the velocity of a satellite, r is its radius of orbit, c is the speed of light, R_L is the luminiferous radius, P is the pressure of the aether, and ρ is the density of the aether. BlackHole.frink in Appendix A computes the gravitational parameter from its luminiferous radius and the pressure of the aether.

It wasn't obvious from the beginning that Eq 4-2 had anything to do with surface tension. The observation that $\rho\varkappa = PR_L$ eventually led the author to realize that surface tension must be important, and must result from the equilibrium at the interface between flowing aether and swirling matter.

4.3 surface tension

Multiplying both ends of Eq 4-2 by density makes it clear that surface tension (Υ) is

Eq 4-3
$$\Upsilon = PR_L = \rho\varkappa,$$

where P is the pressure of the aether, R_L is the luminiferous radius of the black hole, ρ is the density of the aether, and $\varkappa$ is the gravitational parameter of the black hole.

4.4 gravitational parameter from surface tension

Dividing both ends of Eq 4-3 by the density of the aether reveals the gravitational parameter ($\varkappa$) of any given black hole is

Eq 4-4
$$\varkappa = \frac{\Upsilon}{\rho},$$

where Υ is its surface tension, and ρ is the density of the aether.

4.5 luminiferous radius due to surface tension

Another consequence of Eq 4-3 is that the surface tension (Υ) of any given black hole relative to the pressure of the aether (P) determines the luminiferous radius (R_L) of that black hole,

Eq 4-5
$$R_L = \frac{\Upsilon}{P}.$$

Finally, these equations begin to demonstrate that the density and pressure of the aether explain the surface tension of black holes, their gravitational parameters, and their luminiferous radii—in addition to their mass and energy.

4.6 pressure gradient of black hole

Every black hole has a pressure gradient (P) defined by

Eq 4-6
$$P = \frac{\Upsilon}{r},$$

where γ is the surface tension of the black hole, and r is the radius of orbit (or distance from the surface of the black hole).

4.7   pressure is density times velocity squared

Relative to the aether, the pressure (P) of any moving object is always

Eq 4-7 $$P = \rho v^2,$$

where ρ is the density of the aether, and v is the velocity of the object relative to the aether. For an electron in orbit around a nucleus, $P_E = \rho v_E^2$.

    Throughout the pressure gradient, $P = \dfrac{\gamma}{r}$ (Eq 4-6), emanating from the surface tension of any given black hole, Eq 4-7 guarantees that the ρv^2 pressure of each satellite always equals some pressure within the gradient. Consequently, buoyancy causes the satellite to float within the surface tension of the black hole to a level that equalizes the pressure. The satellite must float farther away (lower velocity) to reduce pressure, or closer (higher velocity) to increase pressure.

5.1   charge squared per electron mass

The charge squared per electron mass,

Eq 5-1 $$\chi_E = \frac{q^2}{4\pi\rho\S R_E^3}$$

(2.24245e-9 A^{2}s^2/kg), is critically important to both the electric and magnetic constants, where q is the elementary charge, ρ is the density of the aether, and $\S R_E^3$ is the volume of an electron (so that $\rho\S R_E^3$ is the mass of the electron). That is how the chiFactor method of Electromagnetism.frink in Appendix A computes the charge squared per electron mass.

5.2   electric constant from surface tension

The next few stages explain how the electric constant emerges from surface tension.

5.2a   mass of electron & gravitational parameter of proton

The electric constant is

Eq 5-2a $$\epsilon_0 = \frac{q^2}{4\pi m_E \varkappa_P},$$

where q is the elementary charge, m_E is the mass of the electron, and $\varkappa_P$ is the gravitational parameter of the proton.[63]

---

63 Heffron, Michael. *Ramblings of an Old Man: about the scientific method.* (self-published via Kindle Direct Publishing, 2020). Chapter 7 (page 54).

5.2b   accounting for charge squared per electron mass

Substituting χ_E for the terms identified by Eq 5-1, the electric constant is

Eq 5-2b
$$\epsilon_0 = \frac{\chi_E}{\varkappa_P}.$$

5.2c   substitution

Substituting Eq 4-2 for $\varkappa_P$, that is equivalent to

Eq 5-2c
$$\epsilon_0 = \frac{q^2}{4\pi\rho\S R_E^3 \left(\frac{PR_P}{\rho}\right)}.$$

5.2d   canceling densities to produce electric constant

Compare Eq 5-2c to Eq 4-1c & Eq 5-1. Note that the densities cancel out, which transforms Eq 5-2c into

Eq 5-2d
$$\epsilon_0 = \frac{q^2}{4\pi\S R_E^3 (PR_P)}.$$

Notice that the **pressure of the aether produces the electric constant.** From Eq 5-2, substituting $\frac{\rho}{\gamma_P}$ for $\varkappa_P$ in Eq 5-2b, the electric constant is

Eq 5-2
$$\epsilon_0 = \frac{\rho\chi_E}{\gamma_P} = \frac{\chi_E}{\varkappa_P},$$

where ρ is the density of the aether, χ_E is the charge squared per electron mass found by Eq 5–1, and γ_P is the surface tension of the proton. Notice that the density in the numerator cancels out the density within the charge squared per electron mass (Eq 5-1), thereby making this equation equivalent to Eq 5-2d. Note also that $\frac{\rho}{\gamma_P}$ is the inverse of the gravitational parameter (Eq 5-2), thereby making this equation equivalent to Eq 5-2b. Due to so many different views, it can be easy to overlook the revelation of Eq 5-2d that pressure of the aether dominates the electric constant.

5.3   magnetic constant

The next few stages explain how the magnetic constant emerges from surface tension.

5.3a reciprocal of electric constant

The magnetic constant is the reciprocal of the electric constant divided by the speed of light squared. Canceling densities in Eq 5-2c, then inverting and dividing by the speed of light squared, the magnetic constant is

Eq 5-3a
$$\mu_0 = \frac{4\pi \left(\S R_E^3\right) PR_P}{c^2 q^2}.$$

5.3b substituting

Substituting the density and pressure of the aether for c^2, that becomes

Eq 5-3b
$$\mu_0 = \frac{4\pi \left(\rho\S R_E^3\right) PR_P}{Pq^2}.$$

5.3c canceling pressures to produce magnetic constant

The pressures cancel, so that

Eq 5-3c
$$\mu_0 = \frac{4\pi \left(\rho\S R_E^3\right) R_P}{q^2}.$$

5.3d surface tension becomes apparent

Substituting the charge squared per electron mass (Eq 5-1) for the appropriate terms, that becomes

Eq 5-3d
$$\mu_0 = \frac{R_P}{\chi_E}.$$

Hidden within the charge squared per electron mass, **density of the aether produces the magnetic constant**. Multiplying Eq 5-3d by pressure over pressure demonstrates its derivation from the surface tension of the proton,

Eq 5-3
$$\mu_0 = \frac{\gamma_P}{P\chi_E} = \frac{R_P}{\chi_E}.$$

5.4 electric fields

Advanced readers may wish to explore the electric **D** and **E** fields, $\mathbf{D} = \epsilon_0 \mathbf{E}$, in terms of

Eq 5-4
$$\epsilon_0 = \frac{\rho\chi_E}{\gamma_P} = \frac{\rho\chi_E}{PR_P} = \frac{\chi_E}{\varkappa_P}.$$

5.5 magnetic fields

Similarly, it may be valuable to explore the magnetic **B** and **H** fields, $\mathbf{B} = \mu_0\mathbf{H}$, in terms of

Eq 5-5
$$\mu_0 = \frac{\gamma_P}{P\chi_E} = \frac{\rho\varkappa_P}{P\chi_E} = \frac{R_P}{\chi_E}.$$

6.1 resonant frequency

The resonant frequency (f_r) of an electronic circuit is

Eq 6-1
$$f_r = \frac{1}{2\pi\sqrt{LC}},$$

where L is inductance and C is capacitance. Thus multiplying an inductance by a capacitance produces seconds squared, so that $\sqrt{LC}$ is the period (in seconds) of one oscillation of the resonant frequency.

6.1a inductance in henries

The units of inductance (henries) are $kg \cdot m^2/s^2A^2$ (joules of energy per ampere squared). The magnetic constant (Eq 5-3, luminiferous radius) is henries per meter (H/m).

6.1b capacitance in farads

The units of capacitance (farads) are $s^4A^2/kg \cdot m^2$ (ampere·seconds squared per joule of energy). The electric constant (Eq 5-2, inverse gravitational parameter) is farads per meter (F/m).

6.1c force in newtons

Looking back at the mass flow per second (kg/m^2), that is the N/m of electromagnetic tension. It seems very similar to the H/m of inductance and the F/m of capacitance. That is because inductance is tension per A^2/m^2 and capacitance is A^2/m^2 per tension, where A/m is magnetic field strength.

6.2 frequency of a black hole

The frequency (f) of a black hole is

Eq 6-2
$$f = \frac{c}{2\pi R_L},$$

where c is the speed of light, and R_L is the luminiferous radius. Multiplying by the denominator, that becomes

Eq 6-2a
$$c = 2\pi R_L f.$$

Substituting $R = \sqrt[3]{\dfrac{m}{\rho\S}}$ (Eq 3-6) for R_L, that becomes

Eq 6-2b
$$c = 2\pi f \sqrt[3]{\frac{m}{\rho\S}}.$$

Cubing both sides of that equation produces

Eq 6-2c
$$c^3 = (2\pi f)^3 \left(\frac{m}{\rho\S}\right).$$

Then, dividing by the cube on the right, that becomes

Eq 6-2d
$$\frac{c^3}{(2\pi f)^3} = \frac{m}{\rho\S}$$

so the mass of the electron is

Eq 6-2e
$$m = \rho\S \left(\frac{c}{2\pi f}\right)^3.$$

After dividing by $\rho\S$ and applying Eq 3-6,

Eq 6-2f
$$R_L = \frac{c}{2\pi f}$$

identifies the true radius of the sphere that surrounds the black hole—within which internal aether particles swirl ever so slightly faster than the speed of light.

6.3   surface tension of light

Regardless of its wavelength (λ), light has a constant surface tension (Υ_λ) of

Eq 6-3
$$\Upsilon_\lambda = \rho \varkappa_\lambda = PR_\lambda,$$

where ρ is the density of the aether, P is the pressure of the aether, $\varkappa_\lambda = \dfrac{\Upsilon_\lambda}{\rho}$ (2.18066e5 m³/s²) is the constant gravitational parameter of light, $R_\lambda = \dfrac{\Upsilon_\lambda}{P}$ (2.42631e-12 m) is the constant luminiferous radius of light, and Υ_λ is 3.89136e21 kg/s² is the surface tension of light. Thus, longitudinal oscillations of light behave like a massive particle traveling at the speed of light. Is it?

6.4   pressure of light

For any given wavelength (λ) of light, the pressure of the light is

Eq 6-4
$$P_\lambda = \frac{\Upsilon_\lambda}{\lambda},$$

where $\Upsilon_\lambda = \lambda P_\lambda = 3.89136e21$ kg/s² is the constant surface tension of light.

    Earlier drafts of this book described interactions between electrons and light in terms of their pressures. Thus, it was obvious early on that

$\lambda P_E = 3.89136e21$ kg/s². It took a while to even consider that $\lambda P_E = \gamma_\lambda$ was the constant surface tension of light.

6.5   pressure of electron

The pressure of an electron is

Eq 6-5
$$P_E = \frac{\gamma_N}{r_E},$$

where $\gamma_N = P_E r_E$ is the surface tension of the nucleus, and r_E is the radius of the electron's orbit (or its distance from the nucleus).

6.6   wavelength of light emitted by electron

The next few steps determine what wavelength of light an electron emits.

6.6a   equal pressures

Interaction between electrons and light equalizes their pressure,

Eq 6-5a
$$P_E = P_\lambda$$

6.6b   merge equations

Merging the right sides of Eq 6-4 & 6-5,

Eq 6-6b
$$\frac{\gamma_N}{r_E} = \frac{\gamma_\lambda}{\lambda}.$$

6.6c   cross multiply

which leads to

Eq 6-6c
$$\lambda \gamma_N = r_E \gamma_\lambda.$$

Note that both sides of that equation are force. Consequently, an electron emits light of wavelength

Eq 6-6
$$\lambda = \frac{\gamma_\lambda}{P_E} = \frac{\gamma_\lambda r_E}{\gamma_N},$$

where γ_λ is the constant surface tension of light, and P_E is the pressure of the electron (where r_E is the radius "of orbit" for the electron that emitted the light, and γ_N is the surface tension of the nucleus). Those parenthetical terms are "per the pressure of the electron" (inverse of Eq 6-5).

6.7   radius of electron's orbit

From the preceding analysis, the radius of an electron after absorbing light is

Eq 6-7
$$r_E = \frac{\gamma_N}{P_\lambda} = \frac{\lambda \gamma_N}{\gamma_\lambda},$$

where γ_N is the surface tension of the nucleus of the atom, and P_λ is the pressure of the light (where λ is the wavelength of the light, and γ_λ is the surface tension of light). Those parenthetical terms are "per the pressure of the light" (inverse of Eq 6-4), which must equal the pressure of the electron (P_E), so

6.8　radius of electron's orbit due to pressure

the pressure of the light is $P_\lambda = \frac{\gamma_\lambda}{\lambda}$ (Eq 6-4), and

Eq 6-8
$$r_E = \frac{\gamma_N}{P_\lambda} = \frac{\gamma_N}{P_E}.$$

8.1　Balmer equation for spectral lines

Balmer discovered the wavelength (λ) of each spectral line of hydrogen could be found by an equation equivalent to

Eq 8-1
$$\lambda_{m<n} = \lambda_{Balmer} \left(\frac{n^2}{n^2 - m^2} \right),$$

where λ_{Balmer} is 3.64507e-7 m, and n is an integer that always exceeds the integer m. It is important to note that $\lambda_{0<1} = \lambda_{Balmer}$ is exactly eight times the wavelength ($\lambda_{Bohr} = 4.55634e\text{-}8$ m) of an electron in orbit at the Bohr radius of the hydrogen atom ($a_0 = 5.29177e\text{-}11$ m). That is $\lambda_{Balmer} = 8\lambda_{Bohr}$. It may also be of interest to note that $r_E = 8a_0 \left(\frac{n^2}{n^2 - m^2} \right)$, where r_E is the radius of the electron's orbit for the specified integers m < n. In other words, $\left(\frac{n^2}{n^2 - m^2} \right) = \frac{\lambda}{8\lambda_{Bohr}} = \frac{r_E}{8a_0}$, where a_0 is the Bohr radius (5.29177e–11 m).

　　After Balmer discovered the m=2 series of spectral lines, other researchers subsequently discovered the following additional series: Lyman series, m=1; Paschen series, m=3; Brackett series, m=4; Pfund series, m=5; Humphreys series, m=6; plus many other unnamed series, where m>6.

　　From Eq 6–3, for any given wavelength (λ) of light, the pressure of the light is $P_\lambda = \frac{\gamma_\lambda}{\lambda}$, where $\gamma_\lambda = \lambda P_\lambda = 3.89136e21$ kg/s^2 is the constant surface tension of light. When that light interacts with an electron, the pressure of the light must equal the pressure of the electron $P_E = \frac{\gamma_N}{r_E} = \rho v^2$, where γ_N is the surface tension of the nucleus, r_E is the radius of the electron's orbit (or its distance from the nucleus), ρ is the density of the aether, and v_E is the velocity of the electron. Also note that $\gamma_\lambda = \rho v^2 r_E = \rho \varkappa_P = P_E r_E$, where $\varkappa_P$ is the gravitational parameter of the proton.

Both pressures are equal to

Eq 8-2
$$P_{m<n} = P_{Balmer} \left(\frac{n^2 - m^2}{n^2} \right),$$

where $P_{Balmer} = 1.06757\text{e}28$ kg/m·s^2 is the pressure of the electron and the light when they interact at the Balmer wavelength.

The pressure of an electron ($P_E = \Upsilon_P/a_0$) orbiting within the surface tension of a proton ($\Upsilon_P = 4.51946\text{e}18$ kg/s^2) at the Bohr radius ($a_0 = 5.29177\text{e-}11$ m) times the wavelength of light ($\lambda = 4.55634\text{e-}8$ m) is the constant surface tension of light ($\Upsilon_\lambda = 3.89136\text{e}21$ kg/s^2).

Chapter 5 found that, for any given wavelength (λ) of light, the pressure (P_λ) of the light is $P_\lambda = \dfrac{\Upsilon_\lambda}{\lambda}$ where $\lambda P_\lambda = \Upsilon_\lambda = 3.89136\text{e}21$ kg/s^2 is the constant surface tension of light. Interaction between electrons and light equalizes their pressure, $\dfrac{\Upsilon_N}{r_E} = \dfrac{\Upsilon_\lambda}{\lambda}$, where Υ_N is the surface tension of the nucleus of the atom, and r_E is the radius of the electron that interacted with the light. That is, an electron emits light of wavelength $\lambda = \dfrac{\Upsilon_\lambda r_E}{\Upsilon_N}$, or its radius of orbit becomes $r_E = \dfrac{\lambda \Upsilon_N}{\Upsilon_\lambda}$ when it absorbs light.

The pressure of an electron ($P_E = \gamma_P/a_0$) orbiting within the surface tension of a proton $\Upsilon_P = 4.51946\text{e}18$ kg/s^2 at the Bohr radius ($a_0 = 5.29177\text{e-}11$ m) times the wavelength of light (λ) is the constant surface tension of light ($\gamma_P = 3.89136\text{e}21$ kg/s^2). That is, $\lambda P_E = \lambda \Upsilon_P/a_0 = \Upsilon_\lambda$ so $\lambda \Upsilon_P = a_0 \Upsilon_\lambda$ or (more generally) $\lambda \Upsilon_P = r_E \Upsilon_\lambda$ (Eq 6-5c).

8.3   energy of electron

It is customary to find the energy of an electron using the equation $E = hf$, where h is the Planck constant and f is the frequency of an interacting light wave; however, it is equally valid to find the pressure of the light (Eq 6-4) times the volume of the electron, which is

Eq 8-3
$$E_E = \left(\frac{\Upsilon_\lambda}{\lambda} \right) \left(\S R_E^3 \right) = P_\lambda \left(\S R_E^3 \right),$$

where Υ_λ is the surface tension of light, λ is the wavelength of light, and R_E is the luminiferous radius of the electron.

8.3a   Planck constant

From that, the Planck constant is

Eq 8-3a
$$h = \left(\frac{\Upsilon_\lambda}{\lambda f} \right) \left(\S R_E^3 \right) = \left(\frac{\Upsilon_\lambda}{c} \right) \left(\S R_E^3 \right),$$

where Υ_λ is the surface tension of light, λ is the wavelength of light, f is the frequency of the light, $\lambda f = c$ is the speed of light, and R_E is the luminiferous radius of the electron. That is, the Planck constant is equal to the pressure of the light times the volume of the electron divided by the frequency of the light (or the volume of the electron times the surface tension of light per the speed of light). Notice that the Planck constant represents the rate of energy transfer (joule seconds, like angular momentum).

8.4   modified Balmer equation

The Balmer equation is equivalent to

Eq 8-4
$$\lambda_{m<n} = \lambda_{Balmer}\left(\frac{n^2}{m^2}\right) \div \left(\frac{n^2}{m^2} - 1\right),$$

which explains why $\dfrac{n}{m}$ corresponds to any given wavelength regardless of the values of n and m.

D Abbreviations, Acronyms, & Definitions

3D   Three dimensional.

γ_λ (Greek letter gamma, sub lambda) Surface tension of light, 3.89136e21 kg/s^2.

γ_P (Greek letter gamma, sub P) Surface tension of a proton, 4.51946e18 kg/s^2.

ϵ_0 (Greek letter lunate epsilon, sub naught) electric constant, permittivity of free space, also known as distributed capacitance of the vacuum–defined precisely as $\epsilon_0 = \dfrac{\rho \chi_E}{\gamma_P}$, where ρ is the density of the aether, χ_E is the charge squared per mass found by Eq 5–1, and γ_P is the surface tension of the proton.

$\varkappa$ (Greek letter Koppa) ancient Greek letter used as substitute for $v^2 r$ where v or r could be confused with another velocity or radius. In terms of aether, $\varkappa = v^2 r = c^2 R_L = \dfrac{P R_L}{\rho}$, where P is the pressure of the aether, ρ is its density, and R_L is the luminiferous radius of the black hole.

$\varkappa_{earth}$ (Koppa sub earth) Gravitational parameter of earth, 3.986004418e14 m^3/s^2.

$\varkappa_\lambda$ (Koppa sub lambda) Gravitational parameter of light, 2.18066e5 m^3/s^2.

μ_0 (Greek letter Mu, sub naught) magnetic constant, permeability of free space, also known as distributed inductance of the vacuum– defined precisely as $\mu_0 = \dfrac{\gamma_P}{P \chi_E}$, where γ_P is the surface tension of the proton, P is the pressure of the aether, and χ_E is the charge squared per mass found by Eq 5–1.

ρ (Greek letter Rho) density of the aether, $\rho = 1.7844886677\mathrm{e}16$ kg/m^3.

χ_E (Greek letter Chi, sub E) charge squared to mass ratio found by Eq 5-1, $\chi_E = \dfrac{q^2}{4\pi\rho(\S R_E^3)} = 2.24245\mathrm{e}{-9}$ A^{2}s^2/kg.

$\S$ shape factor for subatomic particles emerging from the aether, which are best visualized as amorphous blobs. Since those particles often behave like spheres (probably due to surface tension), treating $\S$ as if it's the scalar volume of a sphere ($4\pi/3$) is often a good approximation.

a_0 Bohr radius ($5.29177\mathrm{e}{-11}$ m).

c Speed of light, defined by international agreement as exactly 299,792,458 m/s; however, $c = \sqrt{\dfrac{P}{\rho}}$.

C   Coulomb–equivalent to $6.24151\mathrm{e}18$ elementary charges.

CODATA   COmmittee on DATA for science and technology.

constant   An illusion. Nothing is constant, although we can easily mistake stable variables for constants.

h Planck constant $h = \left(\dfrac{\gamma_\lambda}{c}\right)\left(\S R_E^3\right) \approx 6.62606\mathrm{e}{-34}$ kg·m/s^2, where γ_λ is the surface tension of light, c is the speed of light, and R_E is the luminiferous radius of the electron. Notice that the Planck constant represents the rate of energy transfer, like joule seconds (angular momentum).

kg   Kilogram (about 2.2 pounds).

kilo   Thousand.

km   Kilometer (about 1.609 mile).

luminiferous Transmitting, producing, or yielding light. Luminiferous radius is the distance from a black hole at which a satellite orbits at the speed of light. Luminiferous volume refers to the capacity of the sphere that circumscribes the luminiferous radius.

m Meter (39.37 inches, or about a yard).

m_E Mass of electron, $m_E = \rho \S R^3{}_E$, 9.10938356e-31 kg.

m_P Mass of proton, $m_P = \rho \S R^3{}_P$, 1.672621898e-27 kg.

mega Million.

micro One millionth ($^1/_{1,000,000}$).

milli One thousandth ($^1/_{1000}$).

mm Millimeter (0.03937 inch, about midway between $^1/_{32}$ and $^1/_{16}$ of an inch).

nano One billionth ($^1/_{1,000,000,000}$).

NIST National Institute of Standards Technology.

P Pressure of the aether, 1.60381843154e33 kg/m*s^2.

q Elementary charge (1.6021766208e-19 C).

R_E Luminiferous radius of electron, 2.30124148779e–16 m.

R_P Classical radius of "electron" = luminiferous radius of proton, 2.8179403227e–15 m.

s Second.

SI International System of Units, from French: Système International (d'unités).

v^2r Velocity squared times radius of orbit.

ABOUT THE AUTHOR

According to his wife, Michael Heffron is not the kind of doctor who does anyone any good. She once called him over to her computer to look at an email. She said "Look here, I can buy the same degree you have, and you wasted all of those years in school. Are you sure you're smart enough to be a doctor?" Her question has no easy answer!

A decade before becoming a Doctor of Computer Science, Michael Heffron invented U.S. Patent 5,113,832 (Method for Air Density Compensation of Internal Combustion Engines). That invention, more than any other single event in his life, taught him that the majority opinion (even of professionals) is often wrong and tends to stifle the advancement of knowledge. To thrive, we need a free and open society where it is permissible to ask questions, discuss ideas without ridicule, and where we think things through for ourselves. We also need to be perpetually aware of the flaws in what we think we know!